Konrad
Lorenz

Konrad
Lorenz

Alec Nisbett

J. M. Dent & Sons Ltd. London

Made in Great Britain
at Heffers Printers Ltd., Cambridge
for
J. M. DENT & SONS LTD
Aldine House, Albemarle Street, London

This book is set in 11 on 13pt Baskerville

ISBN 0460 04215 7

Errata
page vi: line 4 for coridae read corvidae
           lines 18 and 19 for Dimitri read Dmitri

Contents

Illustrations

The line drawings on the title-page were originally intended to illustrate 'Animals as a Nuisance' in *King Solomon's Ring* but were never used. 'Adolf's tea-party' (p. 71) was drawn for the same book. *Courtesy Peter Wait of Methuen and K. Lorenz*. 'The Bird with the Silvery Eyes' (p. 31) and 'Babette' (p. 109) were sketched for the author in 1973, *Courtesy K. Lorenz*.

For my daughters
Caroline and Sarah
and my son Guy

Author's Foreword

I offer thanks to all who have helped me: Bayerischer Rundfunk and the BBC, who first gave me the opportunity to meet and observe Lorenz; and then to those who, by awarding me a Glaxo Travelling Fellowship, both encouraged and enabled me to continue my investigations. This would still not have been feasible without the help of my wife Jean who objected no more than was reasonable when I seemed to be doing two or three full-time jobs at once. She herself did a great deal to help put this book together.

My thanks go generally to all those who have given readily of their time—though they all clearly enjoyed talking about Lorenz. Alfred Seitz, in particular, contributed a fund of stories of pre-war Altenberg.

I must also thank Konrad Lorenz himself, who patiently endured long series of questions during the eighteen months or so that this book was in the making—though again it was generally of no pain, for few men can enjoy talking more than Lorenz. Few, for that matter, are more enjoyable to listen to. On occasions when the questions, of necessity, were painful, he answered freely and generously nevertheless.

Special thanks go to those who have read all or part of my manuscript and made valuable suggestions for its improvement. These include my friend and colleague, Nigel Calder, who has approached this book with a knowledge and understanding of what has been going on in the sciences surrounding ethology. Another is Wolfgang Schleidt who worked at Altenberg, Buldern and Seewiesen as a pupil and friend of Lorenz, and who did much to help me order the post-war history.

But despite all the help and my best attempts to check and re-check my facts, some mistakes and misunderstandings will doubtless remain. Unlike much of science, history cannot be repeated and reviewed for the purpose of verification or disproof. It is shadowed and intangible, appearing differently in every eye that observes it. The selection here is mine. If there are errors, these too are mine.

ix

Introduction

In writing this book I have had to rely on many sources, for this is 'Konrad Lorenz Observed' by many people beside me.

How reliable is such a process? It must be recognised that observations and their observers are never as independent as the rules of scientific investigation seem to demand; and even less so when observations and judgment are, as in both Lorenz's science and this account of the man, almost entirely qualitative. Inevitably, questions of subjectivity and bias are raised.

If two people are asked about some particular event of which both should know, they are likely to report it differently; so is the truth that which is common to the two statements? Not if the two reports are part of the same chain of rumour. Is it then more likely to come from the one closer to the start of the chain? Not if the second has more and better background understanding against which to evaluate what is said. So would it be better in all cases to go back to the primary source (in this case usually Lorenz himself) and ask for 'the truth'? The answer, again, is: not always, for the man at the centre of the events described has a special viewpoint of them as well as a special interest in their interpretation. Sometimes his may be the only report available – whether it be on remote events or subjective feelings. But he may remain totally unaware of things happening behind him or while he closes his eyes to sleep; he may also undervalue or disregard those things which, consciously or otherwise, he has no wish to see. Lorenz has a genius for the observation and interpretation of certain types of event. But of men, and especially of himself, he may not be the ideal observer. In this book we shall see the world of Konrad Lorenz from the outside as well as from his own viewpoint: shadows are cast from different angles. By this combination, we may be able to see the man himself more clearly – I hope, in the round.

I have relied heavily on the evidence of Lorenz's associates, cross-checking the stories from one to another where possible. Documentary evidence has been sought where appropriate, but is not always available (except on such matters as scientific papers). Newspaper and other similar reports are taken as secondary sources and treated with caution; nobody likes to spoil a good story, and newspaper reports on Lorenz are, I have found, fallible. Quotations in this text are often from transcripts of conversations, or from letters to the author; where they are from books or papers, this is stated. Adolf Lorenz (Konrad's father) provides an exception to this, for in his autobiography (written in English) he reports the vigorous, direct speech of conversations from long before. In this account I have quoted him directly, but allowance should be made for Adolf's marked capacity for literary licence.

As Konrad Lorenz observes animals, so my role (at least in part) has been to observe the man. As we shall see, Lorenz's background and interests affect the way he works, the way he interprets his results and the development of his ideas. In approaching this book, too, the reader should be aware of the author's own viewpoint, for this must also affect the picture of Lorenz presented. My own attitudes and background are often very different from those of Lorenz. My defence against the charges of subjectivity and bias is to declare them here, so the reader may identify with or discount them as he wishes.

First, I am British while he is Austrian. This may, in fact, have certain advantages, for there are aspects of Lorenz that would hardly be noticed by an Austrian but which the outsider sees at once and so is prompted to analyse and explain them as part of the system that has strongly influenced the man. Beside this must be set the national differences still remaining from the Second World War. Those on the winning side (who are old enough to remember) cannot easily forget Nazism, while those Europeans who were persecuted by the Nazis have good reason never to forget. The losers have a greater interest in looking predominantly to the new, post-war world and forgetting the traumatic Hitlerite period, which makes it dangerous to look too closely into an older man's past. This has produced a discontinuity in German thinking which is not reflected in that of her former opponents. In Lorenz's life much happened at this point of discontinuity.

Another difference lies in language, and also in certain habits of thought. Robert Martin, in his introduction to a selection of Konrad Lorenz's scientific papers, notes the difficulty he had in translating certain words into English. This was partly because, in the progress of the science, some terms had become archaic: but there is more to it than that. Part of the problem seems to be that the German language seeks absolutes and a more precise definition of meaning. The poly-syllabic compound nouns that both amuse and irritate the English schoolboy are an attempt at clear labelling which is not so strongly embedded in English, a language which makes a virtue of short words that can be arranged with great flexibility to convey a wide range of shades of meaning. German philosophy, like the German language itself, tends towards absolutes and ideals, and this permeates German intellectual discussion: philosophy is respectable and the language helps it to be so. British 'home-spun' philosophy, that which permeates common intellectual discussion, is—again, like the language—more flexible, empirical and practical. Lorenz, in fact, exhibits a curious mixture of the Continental philosophical tradition (to which the German responds with enthusiasm) and the empiricism that he developed for himself—which the Englishman finds appealing but the German a little puzzling.

A further factor which might have separated Lorenz from the British observer, and still does from most Americans, lies in a past divergence of scientific interests. The reader will not be surprised to discover that my own sympathies lie with the practical British approach to the study of behaviour against the extremes that have until recently existed on either side. There are differences, too, of academic tradition: the formality and seriousness with which Germans (and with them the Austrians) treat their major public occasions pervades public life: Professor Doctor Doctor Lorenz has lived long in a world where distinguished men are ceremonially addressed by not one but several of their titles. It can emphasise any feeling of exclusion that an English-speaking outsider might feel in a German academic environment.

A final and more personal difference from Lorenz is that of scientific background. His was first and formally medicine; subsequently, and by much of a lifetime's study, it has been the behaviour of animals, ethology. My own initial scientific training was in mathematical physics, about as remote as possible from Lorenz's field—especially in view of his personal approach to his subject, which is resolutely qualitative. But since then I have had opportunities to enter and report from the frontiers of many fields, and this has demanded study of the scientists themselves; for in interpreting their work, I have to some degree revealed the men. That, perhaps, is my best qualification for attempting this book.

At the very least, the history of Konrad Lorenz makes a good story. At other levels it is a case study of the interaction between man and ideas, between man and events, between man and his living environment, and between one man and others. Everyone with whom Lorenz has come into any sort of lasting contact has been changed by him, affected at a personal level and in his work and ideas. His effect on science has extended far beyond the range of personal contact. Since writing most of this book I have had the good fortune to work on a project that has taken me to many parts of the world to meet leading scientists in many disciplines concerned in one way or another with human behaviour. Their studies have reached an exciting stage: it seems that for the first time we can begin to approach the possibility of a unified science of human behaviour. As a byproduct, it has been possible for me to see more clearly the role that Lorenz has played in the flow of events towards that point.

This book was not designed as a survey of Lorenz's science of ethology, the study of animal behaviour, although inevitably we shall in its course pass many of the landmarks of its development. Nor is this in the popular sense 'an animal book'; rather, it is about a man and his work, although by way of that interest it must be about his animals too. Much has already been written by and about Lorenz, and there has been no shortage of films (including my own) to illustrate his work and ideas. The purpose of the present book is to indicate a way through the thicket of what are sometimes densely interwoven, and at others only half-related, ideas that one man has launched in a lifetime's industry. To see best how these fit together demands some understanding of the man himself, and his own personal history: my aim is to provide this in a sympathetic but critical biography.

Chapter 1

Sketches from Life

What do most people know about Konrad Lorenz, and how valid is the impression to be gained of the man and his work from his own popular books and the picture of him created by the media?

In a famous old black-and-white film, Lorenz marches along a dusty village street in Lower Austria, well-built and upright, about six feet tall, his bush of dark hair swept back. He has walked here many times before: he is part of this country and his dress proclaims it—today he is wearing baggy cotton plus-fours, fastened below the knee. Occasionally, his shoulders droop a little as he glances back and down to the row of goslings trailing in single file behind him. The young birds have never known their real mother and have attached themselves to Lorenz instead; they are imprinted on him. If he walks faster, they run to catch up; if he stops, they will gather at his feet, waiting to be shown where next to go; and when they reach fresh young grass, they will wander a little away, but not too far, to peck and nibble at it.

Or Lorenz is in a canoe, patiently watching the goslings on the water. This is not the Danube, whose broad waters flow too fast past his village of Altenberg, but a peaceful backwater. When he paddles a stroke or two, they hurriedly turn and follow, huddling closer to the canoe for the moment. The action in both of these scenes is so simple, so apparently uncomplicated, that we may wonder just where is the science? Lorenz does very little but watch over the animals, with patience and understanding.

Another portrait shows him a quarter of a century later. Lorenz's thick hair has turned silver. This, and the grizzled grey beard, give his face more character than before, and his frame has filled out. His waistline reflects a pleasure in hearty eating, though he is tall and broad enough to appear burly rather than plump. By the small lake at Seewiesen, the institute in Southern Germany where he is the leading light, he sits talking before a camera. Somewhat disturbed by the occasion, the goose on his knee evacuates its bowel over Lorenz's trousers. Talking on, he takes out a white handkerchief and wipes at the mess, returning it afterwards to his pocket. A moment later, he takes the handkerchief out again to wipe his face — there is nothing so special in a little goose-soil to draw his attention from the point he is making. Yet what he describes is a detail of behaviour that his own eye would spot while others would see nothing worthy of note.

A few years later still, his pale grey eyes can still focus, sharp and penetrating, on the object of his interest; but now the lids fall back slightly and the surrounding skin is deeply wrinkled: he is approaching seventy. He has drawn part of a car engine on the blackboard for a group of students who laugh as his right arm moves backward and forward like a piston rod, while he imitates its sound '. . . urssssch-shooo-PUP-pffft . . . ursssch-shooo-PUP-pffft'. This is a lighter moment in a lecture on observing and understanding systems, itself part of an attempt to mould the students' view of the nature of scientific under-standing to his own. It is easy to follow and his audience is absorbed; he is a good teacher, strong in the formulation of easily understood ideas, deliberately exaggerating and simplifying. In any group it is he who speaks most, informally darting from subject to subject, interrup-ting himself at times to comment on something noticed or just thought of. When he moves into a clutch of assistants, students or admirers, they tend to edge away, talk less and listen.

Later still, Lorenz is talking to a Scottish journalist on the sunny terrace of the steeply-gabled Gasthaus Göschelseben, high above Grünau and the valley of the swiftly-flowing Alpine river Alm; this is Austria, to the east of Salzburg. Lorenz is talking of the dangers of degeneracy in overcivilisation. His voice is clear and powerful and his Austrian accent marked, but without the twisted vowel sounds of Vienna (near which he lived as a child); only occasionally is his English difficult to follow. It is a subject about which he feels passionately, and the Scot notes: 'When Konrad Lorenz wants to make a point with some force, he lowers his massive head and jabs at the table with a blunt forefinger. "Look!" he says, "I'm an old man and maybe I'm wrong, but . . . ." ' These words are sufficiently characteristic of the man that the journalist decides to place them at the article's head, a piece about Lorenz himself and his views on the sins of civilised man.

Lorenz received a full measure of television and magazine exposure even before he shared a Nobel Prize, and he is quite frank about his reaction to the presence of a camera: 'I become a buffoon. In real life I shuffle along like this.' And his head and shoulders droop in imitation of a morose and introspective old man. 'But when I see the camera, my back snaps upright, a sparkle comes into my eye and I act like a clown.' Accordingly, he has now become more reticent — or ambivalent — about being filmed and seeks to justify it to himself. The money will help to build his new aquarium at Altenberg; there are important things to be said and it is his duty to say them; and it will help to sell his books, which will contribute to further both objectives. But when he has finally agreed to be filmed, he admits that he enjoys it. He also enjoys his fame, and is particularly happy to receive the admiration of those he can himself respect.

That fame, as distinct from academic honour, has clearly come from his popular books and the reaction to them. Of these, the most attractive is *King Solomon's Ring*, an animal book for children of all ages. Delightfully discursive and full of incident, it describes half a lifetime's experience of animals-about-the-house. To the surprise of those who know that much of his scientific reputation rests on his work with geese, there is little in it about greylags, or any other member of the goose and duck family. Clearly, with these birds there must be a great deal of the man himself missing from the book. The strongest personalities that emerge are those of a jackdaw, Jock — 'Tchock' in the original, from the bird's call — and the subsequent colonies of jackdaws that followed Jock in Lorenz's interest.

The social life of these birds is sketched in both words and pictures with swift, sure strokes. When he writes about the jackdaws, the near-caricature that he uses to describe many of his other animals, and even more the humans (himself above all) is gone. By the very contrast in style, we feel and are quickly brought to share his deep affection for members of a species that most of us, even the self-professed animal lovers, would ordinarily regard as a noisy nuisance. But the next time we meet a chattering cloud of these black creatures, we will doubtless look at them with greater interest, perhaps confessing nonetheless that they are not so lovable as the clearly distinguished individual characters shown to us by Lorenz. And we realise that they are seen as individuals not by any clear differences in appearance but by the way they behave towards each other.

We human beings do the same: we recognise each other as much by the way we walk and talk as by differences of physical appearance. 'All Chinese look the same' to those who know few Chinese: and the Chinese may feel the same about Westerners. In making such a statement, we confess our lack of knowledge of the signals that are used

c

3

within the other group, the often tiny but significant movements or intonations that give added meaning to mere words. From the world of the jackdaw Lorenz draws these signals larger than life so that even we can share his experience of living among them.

Perhaps one thing that differentiates man from other animals is that he can have an affection for them without having to imagine himself one of them. The animals we love create the bridge for us: in their brains are implanted some of the patterns of our behaviour so that they come to see themselves as members of our species; they can produce for us the signals which say that they like or adore us, respect or fear us. They are able to ask us for things they want, or tell us the things they think we ought to know. 'Man's best friend' is the dog because most of the dogs that are bred for man today are remarkably good at thinking themselves people.

What distinguishes Lorenz from most of his fellow men is that he works this trick the other way round: seemingly, he can live with and among jackdaws or greylag geese by thinking himself one of them. He 'becomes' jackdaw or goose— or should I say gander! To imagine Lorenz as other than a leading member of what we assume to be the dominant sex of any species is hard. But this would be misleading, as sexual errors in their choice of partner generally go unseen by animals that are persuaded to form attachments beyond the normal boundaries of their kind.

In *King Solomon's Ring* Lorenz unhesitatingly and unblushingly uses terms which are normally applied to human emotions and motivations in order to describe aspects of animal behaviour. As we shall see, he has an excellent defence for this. But for many readers the anthropomorphism—the seeing of animals in human terms—has an effect that he cannot intend. It confirms the anthropomorphism of those in whom it is already a vice, by suggesting that Lorenz supports their own firmly established attitude. He does not, but they may miss the signals that tell them this. Similarly, a scientist of the more precise and pedantic mould could be alienated by this same seeming anthropomorphism, seeing it as part of a circular argument equating the behaviour of animals to that of man, to be followed (in later books) by an extrapolation from animal back to man which must therefore be invalid. Again, Lorenz has answers.

Both of these reactions to the book are extreme, the views of minorities. *King Solomon's Ring* reaches most of its readers readily and quickly, precisely because it is so easy to read. In fact, it merits careful reading, for there is a great deal more in it than may be picked up on the quick skim that its fluent style permits. After such a glance the impression left of Lorenz himself will be at best limited and at worst that of a genial incarnation of Dr Doolittle who talks to the animals . . . for it is there, in the book, that he does. The title of the German original says explicitly, 'He Talks With the Animals, Birds and Fish'. Even among those who have not read the book, the title alone contributes to an image of the man at second remove.

When called upon to speak or to appear in the public eye, spontaneity, for Lorenz, is more important than all the preparation in the world. And yet, he does prepare carefully for each occasion that is important to him. He is concerned not only for the message he wants to get over, but also for his public image; even so, he can easily make an impression far from the one he intends. Since there are many people who would use him for their own ends, he needs the protection and advice of friends. But, as we shall see, he has ventured beyond their care for him in recent years; this is partly because, while still active, he is now in his seventies and the 'children' of his influence have gone their own ways.

Leave aside his next popular book—*Man Meets Dog*—and look a decade later to *On Aggression*. This was designed to explore the animal roots of man's aggressive instinct, although Lorenz frankly states in this book that his ideas on human aggression have gone far beyond the product of his own direct observational experience. His opponents have attacked his proposition at two basic points—each of which deserves a chapter to itself, so strong has been the reaction yet so important the implications whether Lorenz be right or wrong in the detail.

On Aggression has created a second public face for its writer, one which is the obverse of the first. Here we have the positive, assertive Lorenz. Far more people must have heard of *On Aggression* and the ideas it contains than have read the book itself. Through other popular writers such as Robert Ardrey and Desmond Morris these ideas have reached a wider public, modified sometimes in ways unacceptable to Lorenz. And they have even spread further, to lend respectability to fictional representations of man. Along with a selection of other ideas from the study of animal behaviour, they pervade our culture. The concept of 'territory', the space commanded by a hunting animal, was first used by ornithologists in 1920. It has now been extended and adapted to describe an aspect of man's own behaviour. Not so long

5

ago 'territory', like 'ecology', was a technical term which would have required definition; today both words are widely used. At some point behind these changes in our way of thinking about ourselves, Lorenz is to be seen, no longer a Dr Doolittle, but more like a Victorian father, genial enough with meek and obedient children but stern and forbidding the moment they step out of line.

The different characters that may be glimpsed in his writings, and the very way in which the presence of a film camera or press interviewer may change his behaviour (for better or worse) makes it reasonable to ask just how accurate a picture his public may have of him. At best it is incomplete, if we are to judge from the question of a London television previewer, looking forward with limited enthusiasm to a further hour on the man and his work: 'Haven't we had enough of saintly naturalist Konrad Lorenz?' But this very choice of words indicates a need for a fuller and perhaps different view of him. 'Saintly' implies that this person took the Schweitzer-like visual image for the whole man, and the neutral watchfulness of the camera for uncritical reverence.

There are better words to describe Lorenz. It would be more appropriate (if less reverential) to call him colourful, almost flamboyant; assertive in the manner of his speech and controversial in much that he says, but deeply interesting as an individual and as a man in relation to his concept of science. His is an unfashionable concept these days, lying closer to the science of a hundred years or more ago. Research is a field of human endeavour that, for many younger scientists, seems to have become an end in itself, formalised and routine, a standard career choice beyond high school; while for other young people who see the amorality of the pursuit of knowledge as a new form of immorality or social irresponsibility, the whole of science is suspect. Lorenz himself seeks pure knowledge, but allies to this a powerful morality. The combination repays study.

The distinctions that have rewarded the insights he has given to others include honorary professorships at Münster in 1953 and Munich in 1957. He became a member Pour le Mérite for Science and Arts (Germany) in 1950; a foreign member of the Royal Society in 1964 and a foreign associate of the American National Academy of Sciences in 1966. He has been made a member or honorary member of more societies than it would be reasonable to mention. The honorary degree circuit started with Leeds in 1962 and Basle in 1966, moving then to America (Yale and Chicago), and back to England for Oxford and

6

Birmingham. He has received two Gold Medals and the *Prix Mondial Cino del Duca*. In his native land, he was awarded the Vienna City Prize in 1959, the Austrian Distinction for Science and Art in 1964 and, more parochially, a Paracelsus Ring (worn on the finger) from the town of Villach in 1973.

In 1974 there came one-third of a prize from Sweden—the Nobel Prize. There is no award for biology: this was the Prize for medicine. According to the citation it went to Konrad Lorenz, Nikolaas Tinbergen and Karl von Frisch for their discoveries concerning the 'organisation and elicitation of individual and social behaviour patterns'. But beyond the necessarily narrow terms of the award his fellow scientists would in fact give him credit for achievement on a much wider front. They see him as the initiator of an entire field of biological science. In a strict sense this is an overstatement: we shall look at the antecedents of Lorenz's ideas and methods. But there are rare men who, by taking some (though in this case relatively few) existing ideas and methods and dismissing others, can so enlarge them and enhance their value that what was previously a by-way barely worth a detour becomes a thoroughly worth while field of science in its own right.

To some degree, the observation of animals in something like the way that Lorenz did it had already been established as a proper scientific method. There had been individual studies of behaviour centuries before but, according to Lorenz himself, it was Charles Otis Whitman of the University of Chicago who, in the closing years of the nineteenth century, did more than anyone else to develop the new field. His student Wallace Craig carried Whitman's work forward, but in America their approach went out of fashion, so they can hardly be said to have established a science. Much closer to home, in Berlin, Oscar Heinroth had also made objective studies of animal behaviour.

While Lorenz was still at school, Sir Julian Huxley was putting some new and very real science into British ornithology with his celebrated classic study of the courtship habits of the Great Crested Grebe. Wide-ranging in his interests, Huxley moved from the natural selection of behaviour into other fields, and Lorenz heard nothing of the grebe study (or of Whitman's and Craig's work) until he was established on his own path. Huxley, too, came late to a knowledge of Lorenz's growing contribution, but few would now disagree with the title that Huxley has given him: Konrad Lorenz is the father of modern ethology.

7

To outsiders (those who have studied neither biology nor classical Greek) the honour of that title might be clearer if they knew what ethology was, or how its modern form might differ from the vintage or ancient varieties. Analysis of the origins of the word reveals ethology to be a discourse on character or mime which represents meaningful action without the use of language. With W. H. Thorpe we could call it 'the interpretation of character by the study of gesture', or to take this one stage further, 'the comparative anatomy of gesture'.

Lorenz has his own definition: ethology for him is 'the comparative study of behaviour'. While there are many ethologists who no longer believe the word 'comparative' to be important, to Lorenz it implies his own essential method: as in comparative anatomy or physiology, the similarities and dissimilarities of different species are investigated in the hope of discovering how the process of evolution has made each what it is. Ethology started with Darwin who talked explicitly about the evolution of behaviour; it was promoted and extended by Lorenz to become a recognised field of science. It was to distinguish Lorenz from his precursors that Huxley called him the father of 'modern' ethology.

Today, Lorenz might more appropriately be called the grandfather of the science. Those whom he has directly influenced are spread around the world, and they have trained a third generation of ethologists who have already begun to achieve in their own right. Inevitably, this third generation sees Lorenz's contribution less clearly. Their divergences from Lorenz's original method may be enough to arouse him to vigorous protest, but where he accepts that the work is of its kind good, the protest is muted: he merely observes that the current development of ethology is not quite what he had in mind when he began a half century before. By 1951, Niko Tinbergen had already offered a simpler and broader definition of ethology as 'the objective study of behaviour'.

So what was specially new in Lorenz's approach? A few months before it was demonstrated that ethology could earn a Nobel Prize, Tinbergen described for me some of the specific qualities that distinguished the work of his friend:

> He studied animals for their own sake rather than as convenient subjects for controlled testing in severely restricted laboratory conditions. He restored the status of observation of *complex* events as a valid, respectable, in fact highly sophisticated part of scientific

procedure. In the process he discovered many hitherto unrecognised principles, and opened up many new, or almost totally neglected lines of research. Above all, and in essence, he taught very many to look at behaviour with the eyes of biologists—he made countless people aware of the fact that the behaviour of each species is part of its 'equipment for survival and reproduction'; that it is as much the product of evolution-by-means-of-natural-selection as are, say, the structure of the eye, or the functioning of the digestive organs.

Lorenz has shown us goose and jackdaw communities with a rich and subtle social life, full of dramatic incident and well-rehearsed ritual. But his love and respect for animals are not joined by what he would regard as sloppy sentimentality; he handles his animals easily, almost roughly, defending himself against a gander whose instinct for nest defence has gained ascendancy over tolerance of 'mother' Lorenz by clipping the bird on the head with a rolled magazine. Still unable to reach the nest, Lorenz takes the bird and throws it bodily aside, which is less rough than it looks since the creature will, of course, fly. Lorenz is approaching the nest because he has a job to do: to modify his way of dealing with animals because someone may be watching would not occur to him. Similarly, when a pet bird, excited by the intrusion of a film unit into Lorenz's study, sings loudly, warningly and persistently, Lorenz pauses in mid-monologue to throw a pile of papers at it. As may be expected, the bird simply hops agilely out of the way and flies off to look down quizzically from the top of an open door. 'That silences him for a moment', says Lorenz, and resumes his interrupted sentence. He is a brisk, no-nonsense parent to his animal children and has no patience for the sort of animal lover whom that might offend.

Nor, for example, is he anti-vivisectionist, although he would strongly object to the giving of unnecessary pain, and he indignantly rejects the notion that experimentation is of itself evil: 'If you kill twenty rhesus monkeys, or forty, or a hundred, and save one child, I think you are justified.' Few would disagree, and he is contemptuous of those who do as a matter of principle.

He has had plenty of opportunity to study man: why then did he so resolutely choose to study the lower animals? And what is his justification? To this he replies that anything that can be described as justification is not the original motivation. 'The naive justification is that you like doing it, you love animals and you gloat over them in a

stupid way. If they do not give you that simple pleasure, not even an
Asiatic yogi would have the patience to look at animals for as long as
it is necessary to observe them.' There is an element of aesthetic
pleasure in this. Every good animal observer is an 'amateur' in the
true sense of the word; a lover of organic beauty. To Lorenz, a skein
of wild geese flying towards him at his call is a source of eternal wonder.
He has a model for this attitude in the valedictory remarks of his
anatomy teacher in Vienna, when he retired at the age of seventy-one:
'If you ask me what I have done throughout my life in the fields of
research and teaching, then I must honestly say: I have always done
the things which at the moment I considered the greatest fun.'

Lorenz's original motivation was the sheer joy of working with
animals. But is joy alone sufficient justification for a lifetime's work?
Pressed further, Lorenz would add the word 'duty'. He looks at the
Caribbean angler fish, with its seaweed-like camouflage and a fin ray
that dangles a line with a small worm-like appendage seductively
wriggling before its gobble-ready mouth, and says: 'I make it a duty
to investigate every bizarre creature I can lay my hands on.' Such
animals give clear answers to the question 'What for?' This way he
can look directly at the effect of the selection pressures which have
bred structure and the associated behaviour together. Pressed further
still, he has a more practical justification to offer those who are
interested in the social relevance of science. Sweepingly but emphatically, he says that all science is socially relevant: 'Basic science is
socially relevant because it is that which leads to real and applicable
knowledge. When Benjamin Franklin drew sparks with his finger out
of the strings of his kite, he was playing. But the lightning conductor,
the application of what he found, is socially relevant.' This parable is
offered as his defence of other sciences, not his own. The importance
of studying the sociology of animals is, he feels, perfectly obvious.

The most direct application for his work is that it indicates how we
should treat animals both as individuals and as societies, and how to
care and provide for their world while exploiting it for our own
purposes. This in itself offers raw material for the very complex science
called ecology: the study of natural populations as a fully interacting
system. Man himself is part of the total system, although many traditional ecologists are content to omit the human component from
investigations that are already too complicated to handle comfortably.
In recent years, the term 'ecology' has come into common use with a
changed meaning, as 'the impact of man on his natural environment',

and now carries with it a set of value judgments and proposed courses of action. Would it be better perhaps to differentiate between this 'political' ecology and the original scientific discipline? Whichever ecology we mean, Lorenz's work offers us a valuable method for investigating some of its parts: indeed, he offers us methods more than conclusions.

Lorenz has no hesitation in extending the validity of his methods to the study of man, and this may be his greatest contribution. It is also true that when he carries some of the conclusions of his animal work across to man, he breeds controversy. But as he points out, there is no reason to presume that man's central nervous system is constructed in a way that is basically different from that of higher animals: we have the same anatomical features in our brain, although similar parts of it do different things. We do have greater neural complexity, and human behaviour depends to a large degree on our cultural heritage; but Lorenz insists that beneath these the programme laid out in the genetic code still determines many integrated elements of our behaviour, as it does in animals. Man's aggression, he tells us, is a case in point.

With these views, Lorenz is taking on opponents at a wide range of scientific levels, and his manner becomes defiant and unrepentant – for the attacks upon him have been strong. According to the European Left wing, Lorenz's science seeks to justify man and his institutions in the imperfect (i.e. non-socialist) form in which they are found in Western Europe. In their eyes, he provides philosophic support for the Right. But beyond the politically unacceptable scientific theories on the roots of human behaviour, they see darker and more personal reasons for suspicion. Vigorous Nazi-hunters have criticised Lorenz for his accommodation with the Nazi régime in Hitler's Germany, and the attacks have persisted. They were growing in the years immediately before the Prize, and although disregarded by the Nobel jury they were redoubled with the announcement of the award. More recently, Lorenz sought to identify himself with a cause more attractive to the young, when he leapt upon a table at Munich's Hofbräuhaus to espouse and offer himself as a leader of Germany's ecology movement. In fact, from his book *Civilised Man's Eight Deadly Sins*, it emerges that, in detail, he is saying slightly different things about the issues concerned than others might have chosen to say for themselves. And yet, in his individual approach, he shows as always a special perception of matters which affect us all.

So what picture do we have of Lorenz the man? We see him justly praised as an influence crucial to the development of a particular branch of science—though had he not done this, others would doubtless have filled that role later, perhaps in different ways. We see him as an animal lover but also as a critic of those who love animals more than man. A brilliant innovator, he nevertheless disapproves of some of the modifications (or genuine innovations) of those who follow him. He has a genius for observing detail in the natural world around him, but is a less skilful observer of the political motives of those amongst whom he walks. A skilful populariser, he has introduced his insights to a wide readership and induced many to follow in the footsteps of his own followers, but may seem anthropomorphic to some. We see a social and environmental reformer—but one who still seems in some ways authoritarian and paternalistic. There is also the scientific administrator who has set up better and better institutes but who finally is no administrator at all. We have still to meet him as the philosopher of science whose theory rationalises his own approach to the objects of his study.

We may see as many aspects of Lorenz as he sees sins in civilised man; and each light shines against its own shadow. He often avows his optimism, but there are just as often strong shades of pessimism to many of his pronouncements. No single label defines him, nor is it yet clear how the many parts of the man add up to a whole. But we now have an outline to his form in the jigsaw puzzle, and into this we can begin to fit some of the missing pieces.

Chapter 2

Father and Son

There are European cities where the past lives: the bustling activity of the present, branching and thrusting vigorously in new directions, draws sustenance from the roots and so gives them renewed purpose. Vienna is not like this; today its past sleeps and the headquarters of the Austro-Hungarian Empire lies almost still. Austria has settled for a third-ranking world position and is making a reasonably good best of it. The adjustments to that, a part of the backdrop to Lorenz's own lifetime, were twice traumatic to the body of the country that he grew up in and loves.

In the Austria of his childhood, before and well into the First World War, all this lay in the future. Vienna was one of the world's great capitals, though already sleek and poised for military and diplomatic decay. In the scientific and cultural fields it remained a bustling metropolis of ideas and influences whose atmosphere drifted over the young Konrad. But the city was never more than his second home. He spent his childhood and the most productive years of his early maturity in the nearby riverside village of Altenberg, watching the Danube's annual change from the ochres of spring and summer to the blue of winter. Now, towards the end of his life, he has returned there to recreate for study and pleasure some of the influences of his happy childhood.

Vienna lies cupped between the Danube as it flows south-eastward, and an arc of hills to the west and north. From these hills further westward again lies the Wienerwald, the Vienna Woods, in truth a substantial forest which is still scarcely developed, except for the scars of the autobahn. One branch of the forest hooks round behind the hills to the north of Vienna, probing down to the gently curving Danube as it skirts the wooded slopes. Here, between the Wienerwald and the river's fast flowing waters, lies the long, thin double village of Altenberg-Greifenstein. Greifenstein once boasted a castle, brooding on the hillside, to command the river. Today, only ruins remain to

overlook the water and the flat fertile farmland of the opposing shore –
where sits Burg Kreuzenstein, another castle dominating its own
landscape. Altenberg and Greifenstein have no way across the river
and only the narrowest strip of farmland between Altenberg and the
next village, S. Andrä. The road that curves with the Danube to pass
through Altenberg remains narrow and awkward even now, and the
railway line that lies between road and river allows an alternative
form of halting progress to and from Vienna, though it also tends to
cut forest and villages off from the water.

If you search for Lorenz's Altenberg with a gazetteer of Austria you
may find yourself in quite the wrong place, for it is omitted from guide
books which do, however, mention Alten*burg*, a town that boasts an
abbey. Alten*berg* today has little to distinguish it beyond the presence
of its own Nobel prizewinner. But in the year of Konrad's birth,
Altenberg had two, if minor, claims to fame: one was its own mock
Moorish 'castle', Schloss Altenberg, and the other, just along the
street, the house of an orthopaedic surgeon who had pioneered a new
hip-joint operation and who, it was said, had charged a wealthy
Chicago stockyard king a million dollars to cross the Atlantic and set
the bones of the American's congenitally deformed daughter. The
doctor, a distinguished and distinguished-looking man of forty-nine,
was Professor Adolf Lorenz, Konrad's father.

Through the middle of Altenberg—if any line crossing such a
straggle could be said to go through the middle—a small stream runs
down from the forest. Beside this lies the lane which now bears the
name 'Adolf Lorenz Gasse', for it was here that Konrad's father had
decided to build the grand new 'Lorenz Hall' as the culminating
realisation of his childhood dream to become 'ein grosser Herr', a
grand gentleman. Adolf's influence on (or interaction with) his younger
son was to be strong, and this in turn was moulded by his own remarkable history.

Adolf Lorenz was born at a country town in Silesia, then part of
Austria. The son of a harness-maker, he was poor and ill-fed (in part
because of his own aversion to meat) but healthy. The flaxen-haired
boy ran barefoot during the summer months, and his hardened feet
were unhurt by the autumn stubble in the fields; in the winter he was
allowed cheap shoes though without benefit of stockings against the
bitter cold. Later in life he never grew to like the winter and whenever
possible chose that time of year to travel in warmer climates.

At the age of seven, while rummaging in the attic, Adolf found a small, soft ball of fabric which proved to be a left-handed black glove, a remnant, perhaps, of some ancient funeral. Trying it on, he showed the gloved hand to his mother; now, he told her proudly, he was 'a real big gentleman'. His mother smiled down at him and said, 'To be a great gentleman you must have two gloves. Go and find the other'.

This image recurred to him often in later years, and that metaphorical search for the second glove occupied much of his life. At the age of eighty-one he was to subtitle his autobiography *The Search for a Lost Glove.* In his formative years, by a mixture of diligence, a good mind and pure chance, he gained the education that would lead him to become an anatomist, then a surgeon, and, at thirty, to his first disaster—for his incipient success coincided with the introduction of carbolic antisepsis, and the carbolic made raw meat of his sensitive hands. With the promise of a brilliant career ruined he had to pioneer new medical techniques, at least in part to overcome his own disability and to resume his search for the second glove that had been snatched away from him at the moment he felt it to be within his grasp.

Adolf married his capable young assistant, Emma Lecher, a young woman 'of good family' but with no money of her own. It was their seven-day honeymoon at her parents' country cottage that first carried him to Altenberg. There, in that brief respite from Adolf's efforts to build himself a new career, Emma drew his attention to the garden sloping up from a ramshackle peasant house on the other side of the lane. When Adolf stood there, he found he could look out to a magnificent view over the Danube to Mount Oetscher, the last outpost of the Alps; and west to the fertile farmland of the Tullnerfeld. The peasant was in debt and wanted to sell, so Emma tempted her husband. 'We are living from hand to mouth', Adolf told her, and seemed less enthusiastic than she.

The following week they were back at work in Adolf's new orthopaedic practice, which still involved him in open operations, in preparation for which he used alcohol in place of carbolic. An orthopaedic surgeon's most difficult and unrewarding case at that time was a congenital dislocation of the hip; it was the commonest deformity to be seen in children and, for reasons unknown, appeared mostly in girls. Despite his skill, open surgery had poor results, involving as it did the removal of much of the inner surface of the bony socket, the very part from which new material might otherwise have grown. Adolf's great advance was to find 'bloodless' ways by which the ball could be brought to stay against the immature socket; to demonstrate that over a period of months it would then grow to cup the ball, and further, to show that the necessary long period in a fixed position would not lead to a rigid joint. When the treatment was complete, the child could, for the first time in her life, walk both easily and upright.

As an increasingly successful surgeon, he was able to scrape together enough money to buy out the peasant from his coveted vantage point at Altenberg for the price of the peasant's debts: it was now Adolf and not Emma who was all for owning the garden and from that moment it was his wife who repeatedly advised caution as she came to realise that her husband had reckless enthusiasm enough for two. He was soon buying up extra parcels of land to broaden the boundaries of his domain. He repaired and added a second floor to the peasant house, using all his hard-earned money and more; then, falling sick, he dreamed of a roof tiled with banknotes. But it made a suitable 'father-house' for Emma and their first son, Albert. The structure centred on one vast room, a kitchen; the dining-room was above. As the money from his practice continued to roll in, he became solvent again, and began collecting objects of an artistic quality that appealed to his eye: showy but decorative paintings, Roman marble statues, and the marble balustrades of dismantled Viennese bridges. Emma shook her head in exasperation at this crazy husband: approaching fifty, his beard grey all over, he should be thinking of providing for his old age.

His reputation spread, until one day he was summoned to treat a case in America, a girl born in 1896 with congenital hip dislocation. Adolf's immediate reaction was to say no; how could he leave all his other patients in Vienna so long? But Emma was more shrewd: name a huge fee, she told him, and let the American's response decide for him. The patient was the daughter of Chicago meatpacker and stock-yard king, J. Ogden Armour, who could well afford the money, so the fee—the like of a princess's ransom—was agreed. By following his wife's advice Adolf made his fortune.

In the days before film stars could steal the headlines, Adolf Lorenz became a celebrity, known and immediately recognised wherever he went. With a fortune in his pocket, he could also afford to be generous with demonstrations of his method, and often worked 'regardless of fee'. He complained cheerfully that he could never escape the Press. Once, in a railway sleeping car, a reporter pulled on his leg to wake him and ask was it true that he had been paid a million dollars for his operation? Lorenz considered that one leg pulled was good for another: 'Two million', he replied solemly. The true size of the fee is hinted at in his autobiography, but not explicitly stated; and he could not resist embroidering a good story. Even so, according to Konrad, that whole American visit brought in close to a million dollars.

This same book clearly reveals that, in truth, Adolf had always been something of a snob. He had earlier delighted in hob-nobbing with members of the European aristocracy, and now he could enter the houses of the money-privileged of America. Looking around him within these impressive homes, he determined that in this, at least, he could compete. With both the means and mental plan for the Grand Hall at Altenberg, he would now realise his dream. He would be the envy of his colleagues in Vienna, where he still felt he was appreciated less than his worth. Although for some years he had been a 'Professor Extraordinary', the principal out-of-the-ordinary feature of the post was the total lack of any salary to go with it. But unhampered now by lack of money he could create a Hall that would impress even his new American friends. And so it was that he embarked upon a scheme that mixed baroque, art nouveau, and American megalomania. The chosen architect, already a little unstable as he set out to give body to his client's brainchild, ended his life in an asylum for the insane.

Called upon to return to Chicago for a follow-up consultation, Adolf went with far less hesitation than before. But while he was in Chicago this second time he received a message that brought him up short: his wife was pregnant once again. He considered the matter: Emma was forty-two, beyond the age when, in Adolf's medical opinion, she should be bearing a child. A premature birth was possible, even an abortion—perhaps the latter was preferable? He had seen enough of weakling children born to older mothers to have no wish to be the father of one himself. And yet, setting aside all medical and even rational considerations, perhaps there was something in the bracing air of this new country that had given the middle-aged man the vigour of the young. So he thought of the new arrival as 'the American' – in German '*der* Amerikaner', for surely this would be a male child. He finally determined that the best action would be to let things take their natural course, whatever this might be, to take all proper care of the infant, but with no incubator as an artificial womb if it were premature. He wrote, 'The newborn child must be fit to stand the extrauterine life or it had better die'. Reflecting again though . . . if it came to this, would he still have the resolution?

He sailed for Europe in July 1903, to pregnant wife, first son, and to the building of 'Lorenz Hall or bust'. Emma had to be satisfied that at least part of the fortune went into the unassailable security of Austrian State Bonds; only a part was wasted on this extraordinary construction. On the completion of Lorenz Hall, Adolf chose the words to be inscribed over the door: *Consider as gain whatever chance may bring*—and when the house was ready to receive its first guest, that first arrival was Konrad. He was born on 7 November 1903.

The boy-baby was frail and thin, as his brother Albert had been almost twenty years before; it was a hard birth but both child and mother survived. 'As a reward' (Adolf's words) 'she was given five years' furlough from her work as assistant to her husband, so that she could attend her little son in their "Green Paradise".'

The house that was to be Konrad's family home throughout his life is focussed on one vast central room—in truth, a Hall. In the satisfied words of its designer, it was 'a sight to be seen'. Perhaps only thirty feet high, it seems more as the eye travels upward to the ceiling medallion with its twenty-feet long, baroque-style allegory, *The Victory of Peace over War*, commissioned especially for that spot by Adolf. Along one wall rises a massive wooden staircase, turning into a broad landing that bridges before a tall, bowed wall of windows. High along the third and fourth walls is a recessed balcony, arcaded with pillars and giving access to rooms abruptly small by comparison with the hall but cosy enough as bedrooms and the family dining-room. Konrad now sleeps in the room designated, in the original plan, as the nursery that would welcome him as a baby.

At the furthest corner of this ascending, circular tour, a concealed door opens to reveal a staircase winding to the attics above. At the very end of the last corridor is the tiny corner room once occupied by the youthful Konrad and some of his animals. Here, at the age of ten, he kept his first pet bird (in a cubby-hole in the wall beside a high window) and next door in the true attic he later established his jackdaw colony which extended dominion over much of the roof itself. Many years later, the accumulated droppings were sufficient to rot the massive timbers and almost literally bring the house down.

Back in the hall below is a vast open fireplace to the left of which hangs the darkened *Judgment of Diana* that Adolf had found in Vienna. Around the room he arranged a mélange of commissioned pieces portraying, without too much artistic talent, the lusty, busty mythology of the rococo. Above the stairs is *The Four Ages of Man*, of which Childhood is centered on a cherubic group around a pretty young girl in a chariot drawn by small winged angels. The face is that of the

American 'princess' cured by Adolf, little Lolita Armour, whose father's money had built the house. Away behind her back and forgotten by the carefree group, lies an infant on his belly in the grass, his face distorted by weeping—'the document of a naughty boy', wrote Adolf. This unflattering portrait is of his son Konrad. And to the far right is Adolf himself, not as he was then but as he expected to become—in *Old Age*.

The whole effect of the house today, both indoors and out, is softened by seventy years of maturity and use. Ivy grows on the garden walls and over the lower, older wing of the house almost to the windvane, in which is pierced the date 1903. It well befits Konrad's own style and maturity, although even he can be dwarfed by the great central hall. Communication is still a problem, so that the simplest way to locate his wife is not to search but to shout. 'Maidy', he bellows, in a roar that echoes through hall, house and grounds. Short for Mädchen, 'Maidy' is his wife; slender, neat, grey-haired and intelligent, she is three years Lorenz's senior, with a caution to complement his impetuosity.

From Konrad's birth, for over thirty-five years the house had been alive with family, growing children and animals. In one sense it was a hangover from the Victorian style of household, with Adolf the father dominant and respected, but this was compensated for by an extra-ordinary degree of tolerance on the part of both parents for the unusual hobby of their younger child. At first, pet dogs were not permitted, for at that time microbes had recently been discovered, Vienna being one of the settings for the medical detective story that disclosed their existence. His mother was a victim of the exaggerated fear that protected children from this new menace not only by boiling the vitamins out of their milk but also by keeping them from dirty, dangerous, germ-laden dogs. So, instead, Konrad collected river fish and pond crustaceans: these at least were containable.

But year by year, almost month by month, the animal population of Altenberg grew, and his old nanny, Resi Führinger, who had 'green fingers for animals' taught him how to look after them. A gift crocodile, finding his terrarium too cool, soon had to go: it was this that was swapped for Konrad's first dog. Named Kroki in honour of its predecessor, this creature also had an appropriately similar shape. It was a dull dog, a dachshund, but its virtue was that the boy was allowed to keep it. Then there were birds, some of which caused relatively little trouble—until the day when their droppings on the soft furniture proved to contain the indelible purple dye from a feast of black-currants. In later years, to protect the house from invasion by his ravens and cockatoos someone could always leap up to close a window. The untameable greylag geese could (with difficulty) be shooed away if they wandered into the wrong place at the wrong time, but a capacity for house-training seems to have been a matter of little importance to the young animal collector. His favourite illustration of his parents' tolerance for his animals (and so for the escapades of the boy himself) is the story of the lemur he bought while still at high school.

Konrad's Madagascar lemur was not a small animal: its tail alone was some fifteen inches long. He bought it in Vienna, carried it in a box to Altenberg and let it loose in the tea-room where it at once demonstrated its lack of house training and from time to time broke things. It continued to do both for some sixteen years, but more embarrassing to the family was the way it would importune guests who smoked. At the smell of the cigarette it would jump on the visitor's shoulder, work itself into a frenzy of excitement, then grab the cigarette and run. It carried this as it would an insect, by folding up one leg and tucking the still smouldering cigarette into its groin: it galloped away three-legged, pursued by the entire Lorenz household who were anxious to save the animal from getting burned, leaving the visitor amazed and puzzled at the extraordinary antics of both family and animal.

As each new, and possibly more destructive, pet arrived in the house Konrad's parents merely shook their heads or sighed resignedly. He had started at the age of six with a duckling. He had one, and his friend Gretl had the other of the two they obtained, newly hatched, from a neighbour. In those days the Danube marsh spread several hundred yards back inshore from its banks, and the dense thickets of American golden rod, willow, reed and scrub, interspersed with narrow tracks and backwaters, all provided a perfect playground. Here Konrad and Gretl ran and splashed around, pretending to be ducks, and the hatchlings followed them as readily as if they really were. The two children, quite naturally and without thought, did just what the professional goose students at Lorenz's institute were still doing over sixty years later: to use the proper technical term, the ducklings were 'imprinted' on them. 'What we didn't notice', said Lorenz at seventy, 'is that I got imprinted on ducks in the process. I still am, you know. And I contend that in many cases a lifelong endeavour is fixed by one decisive experience in early youth. And that, after all, is the essence of imprinting.'

Imprinting—later a cornerstone of his scientific method—was a simple and unquestioned fact of life: it would be long before he realised that there was anything special about it. But only a few years later, when he was nearly ten, he fell under the influence of the concept that has consciously dominated his whole life.

Konrad had picked up and read a book called *Die Schöpfungstage* (literally 'The days of the Creation') by Wilhelm Bölsche, a popular writer who had already introduced a whole German-speaking generation to the evolutionary theory of Charles Darwin. To the boy, this revelation illuminated the whole of living nature, lending system and order to its otherwise bewildering diversity. Konrad devoured Bölsche; then to satisfy his craving for more knowledge and understanding he searched out every scrap of further information that he could glean from any source.

He recalls vividly a particular day when he took a walk through the forest with his father, and began enthusiastically explaining evolution to Adolf Lorenz. 'I was a great prattler as a child', Konrad told me, 'he didn't silence me, and this was slightly disappointing. *And then I realised that he had known all along!*'

At this moment, there boiled up in the child a deep resentment that something so important should have been kept from him. It was one of those moments that remains brilliant in every detail of the memory: in his seventies, his mind's eye can still vividly see the path through the forest and the point on it where the conversation occurred. The boy decided at once to study evolution and to become a palaeontologist. Dinosaurs joined in his garden games: in their play, Konrad and Gretl became iguanodons.

The girl Gretl (short for Margarethe) is the same 'Maidy' we have already met. Through a busy lifetime of varying fortunes and ever-developing interests, Konrad Lorenz has been buttressed by a single, stable marriage. He was fortunate in his choice of life partner—or, as they delightedly told me the story, *her* choice. A friend of Gretl's, a romantically inclined lady, had been describing how she had first found her own man. She was in an anatomical dissecting room and a young instructor came in, tall, blond and handsome. She took one look and knew instantly: 'That's him!' Having told her story, she turned and asked how Gretl had met Konrad. Gretl considered for a moment: 'Well, I was sitting there in my pram, and another pram drove by. In it there was a fat ugly baby, and at once I knew — that's him!' Lorenz roared with laughter at the recollection of that conversation; Gretl smiled at the retelling and glanced to him, pleased.

Margarethe Gebhardt was, if not the girl next door, close enough to it. Her father was a market gardener in S. Andrä, the next village to the west in the same sweeping bow of the Danube; and as children she and Konrad spent a lot of time together. He was so sure of his girl and the qualities that made her as good as the next lad, that he boasted to other boys that she would walk round the rim of the cistern that watered her father's nursery garden. The cistern is still there today. To provide a good head of water it stands on its tower high above the ground, and the circular rim is narrow. To make good his boast, the young Gretl did not hesitate but in front of the admiring crowd quickly climbed the supporting ironwork, pulled herself up the outside of the tank, stood and walked round the rim. The performance probably enhanced Konrad's prestige — to have such a girlfriend! — as much as hers.

That she was the daughter of a market gardener had another minor value: a mole could be caught to order. This small creature immediately demonstrated to the young Lorenz its incredible appetite for earthworms, of which it consumed more than its own body weight in a day. It was fascinating to watch the speed and manner of disappearance of the black, glossy animal under and through the ground, to locate, by smelling from below, the scattered wormy treasure that Konrad laid out on the surface of its terrarium. But for once a pet defeated the boy: it was never tamed and he soon tired of the spadework necessary to keep the animal in food. He freed it in the garden. Moles and their insectivorous relations are not recommended as pets.

Despite the special interests which occupied much of his time he had a good range of friends of about the same age. From the period spent as a member of a 'wild band' ranging in age from ten to sixteen, he can still remember vividly his admiration of and desire for approval from their just but strict, uncontested leader. His feelings towards this boy who was four years his senior matched the respect in which he has held teachers and special friends in his later life. It was something that he could remember many years later when thinking about the biological basis for the gap between generations.

There was no more than a modest range of books in his parents' home. 'Surgeons are not usually very literate', says Konrad. In this quality his father was easily outshone by the future father-in-law who had more books and a remarkable range of interests which, like the produce of his gardens, was entirely home grown. Lorenz developed a hearty admiration for the self-taught man. Later, much of his own learning of philosophy, and even his knowledge of the scientific literature of animal behaviour, came by self-teaching.

Photographs of the boy show that, at nine, he was an engaging, mischievous, grinning imp; at eleven, a well-dressed, casually superior young man, the master of his environment. Both portraits told the truth, and still do.

Konrad had long been a fisherman with net and jam-jar. For the net he used bent wire and any suitable scrap of fabric. A year before the first ducklings, he already had Danube fish in his aquarium. Fish fanciers used to the brightly coloured miniature fish of today's pet-shop aquaria might find these river species dull and over-similar, interesting (at lengths of some two to four inches) not even for their size. The patient observer is occasionally rewarded by a glint of iridescent silver as scales catch the light. Konrad was, and remains, captivated by these small and apparently undistinguished fish and still has an aquarium containing them.

In freshwater ponds around Altenberg, the nine-year-old boy also found waterfleas, minute crustaceans that required first a magnifying glass and then an inexpensive microscope to reveal their form—and with it more and more of the wonders of the freshwater pond. In *King Solomon's Ring*, Lorenz makes of this a turning point in his life: 'Therewith my fate was sealed; for he who has once seen the intimate beauty of nature cannot tear himself away from it again. He must become either a poet or a naturalist and, if his eyes are good and his powers of observation sharp enough, he may well become both.'

There is a saying in biology that 'ontogeny recapitulates phylogeny', meaning that the form of a foetus and the development of its internal organs as it grows in the womb follow the same line of development that the species itself took as it evolved through history. It should be used as a first-guessing tool rather than as a real indication of all that has gone before. This proposition offers partial explanations: for example, the tailed newt-like stage which the human foetus goes through has been held to suggest amphibian forebears. Lorenz takes this game a stage further, suggesting that the development of individual scientists tends to recapitulate the progress of their chosen fields of endeavour. Most sciences, he observes, go through 'collecting' phases, as Lorenz himself did. The search for systematic order, for scientific law, comes later.

To the scientific attitude he inherited from his parents and developed strongly for himself were added the ethical values absorbed from family and friends. As he described it, he had 'the tremendous good luck to be brought up in two entirely irreligious families' — which he immediately qualified by saying they were 'highly moral and very spiritual'.

Adolf had gone, at the age of eleven, to a Roman Catholic monastery school, but more than Catholicism, he carried away the memory of persecution by a junior monk (on the grounds of the boy's poverty, he believed). The family that he later headed remained nominally Catholic. Gretl's parents were theoretically Protestant. In the Austria of the early twentieth century, religion had become liberal — so much so that it was a Catholic priest who was free-thinking enough to continue Konrad's education in Darwinism, a concept of such beauty and rightness that it had a far greater impact on the boy than any from the realms of theology. When he was almost six, his formal education began at a private elementary school, financed by a wealthy Viennese master-baker called Mendel and managed by Konrad's teacher aunt. By the age of eleven, he had started at the Schotten-gymnasium, one of the best high schools in Vienna, taking at first only chemistry, physics and natural history. By the fourth year he was there full time, imbibing Shakespeare, Homer and the humanities, the usual elements of a broad, liberal education, in addition to the science (and particularly the biology) at which he already excelled. This pleased Adolf, for it would lead in its due and proper course to a career in medicine. The younger son, like his older brother Albert, was expected to build on that which was given to him: any other course would be unthinkable. In fact, Albert followed precisely in his father's footsteps to become a promising orthopaedic surgeon. He was to survive two wars as a doctor, and four marriages, to die in his eighties; he was the white sheep of the family and, as Adolf described him, 'a nice healthy boy . . . a reincarnation of my own father', whom Adolf the son had properly respected.

While Konrad was at high school, the First World War came slowly but inexorably to Altenberg, and as transport became more difficult the twelve miles to Vienna stretched themselves out interminably. Trains were cancelled and the line through the village reserved exclusively for military use. As Adolf's autobiography insistently and graphically describes, there was no fuel for the car, no light at night, and no water, since this depended on a petrol-driven pump. And there was no coal, no carriage, nor food in the market, so the family moved for the duration to their town flat in Vienna's Rathausstrasse, close to the University. There, conditions were even worse—now without even the consolation of the natural beauty and peace of the countryside. While the family was away, the house at Altenberg was plundered three times. On the third occasion the thieves were caught, and one of them was a one-armed war invalid. In Vienna, the plaintive cry of 'how much longer?' gave way to 'what are we to live on?'; but tramping the streets to find a shop with food wore away unobtainable shoe leather. Adolf tried to grow his own tobacco but gave up in disgust at the results; fortunately, he did better with potatoes. In the comradeship of this new poverty, the patients and ex-patients of his ward gave of what they had, to be shared by those in greater need. But of all his father's wartime miseries, Konrad remembers little—except (perhaps in the peace that followed) carving into a big cube of corned beef. That was 'the only clear memory that we were slightly deprived'.

The peace had left Austria dismembered by treaties that, among other indignities, gave Adolf's Silesian birthplace to Czechoslovakia. He was appalled by the injustice of his country's treatment by the Allies and thought bleakly that perhaps they had failed to realise that the Austrian Army, though itself defeated, had stopped the sweeping wave of Russian Bolshevism. This peace of defeat was followed by the hyperinflation recalled by Konrad as a curiosity rather than as a family disaster. 'One million' was a handy, accepted concept. Then, he remembers, came the introduction of the Austrian schilling, a silver piece that for convenience was exactly the same size and weight as the ten-thousand-crown coin it replaced. A calculation of the value of Adolf's carefully invested fortune showed it had diminished to one fourteen-thousandth of its original value, 'not enough for a family breakfast!' It was a loss that impressed Adolf a great deal more than his son, for the house of Konrad's summers, the product of Adolf's supposed extravagance, remained.

Then, out of the blue once again, Adolf received an invitation to the United States from a committee for the relief of German and Austrian children. This, he hoped, might do good not only for the children of Vienna, but might also help to restore his fortunes—which it eventually did, and to something near its former proportions. But in postwar America, Adolf was now a controversial figure. Admired by some, he was identified by others with 'the Hun'. Re-establishing himself professionally was a painfully slow business, until, at the age of seventy, he was seeing patients again.

By 1922, he turned his attention to the future of his younger son. Quite simply, it was time for Konrad to put away childish things, to set aside his animals, and become a doctor like his father and elder brother. And of his childhood friends he would marry, if it could be arranged, the rich daughter of Schloss Altenberg. The relationship between father and son was close: the traditional ties between generations had not been eroded to the degree that Konrad Lorenz claims we see today. When he thinks of the generation gap, which has widened to a chasm in his lifetime, he talks of the relationship of Abraham and Isaac and, in the same breath, of that between his father and himself. In either case, that the son should defy his father was almost unthinkable, so as he defended his intention to follow his own course, the conflict within him must have been powerful. Konrad has never shown much respect for remote authority; but here the pressure was close and personal, and his father did not give way: the medical man looked down on other sciences; zoology, indeed, was hardly to be taken seriously.

But the most urgent need of the moment was to disentangle his son from Gretl. In any case, Konrad was obviously unhappy: in his final year as a schoolboy he was eighteen and she twenty-one. At that time of life the gap in their ages yawned its widest. She was an attractive and capable young woman who had now left school (before completing her studies) to look after her mother and much younger sister, which was necessary because the mother had suffered a complete breakdown when two of Gretl's brothers were killed during the war. To break the liaison once and for all, Adolf sent his son across an ocean to New York where he was enrolled at Columbia University.

While contemplating his misfortunes, Konrad went fishing in Long Island Sound. He expected that instructors at the Zoological Institute would be able to tell him the names of the species that he caught, but much to his disgust, they could not. With his protozoan, crustacean, mollusc or bird he was always directed to the same man, a tall, thin gentleman of sixty, with a goatee beard that made him look like Abraham Lincoln. He was pleasant and always helpful to the young student, but Konrad could not help noticing that he had one curious idiosyncrasy—he seemed fascinated by fruit flies. He had not just one colony, but tier upon tier of little bottles of the buzzing insects, attended by a white-haired lady who watched over their lives and lineage. Eventually Konrad screwed up his courage to ask what all those funny little flies were for. 'It is one of the major prides of my life', he reminisces, 'that I saw my first chromosome in the microscope of Thomas Hunt Morgan — the father of modern genetics'.

But for the most part Konrad felt he was wasting time in New York. When he returned home there would be a lot to catch up on because the three pre-medical years at Columbia would not be accepted at the University of Vienna. And what of Gretl, twenty-one, attractive and free? It was actively dangerous to remain in New York for a moment longer than he must.

His father had given him five dollars a day to live on; he managed on a dollar fifty. Towards the end of the term, he added the difference to a part-refund of his University tuition fees and bought a steamship ticket for Europe. Although the unsatisfactory nature of the course gave the student a good justification for leaving, Adolf was furious. Konrad left America just before Christmas 1922, bearing (as he recalls) the curse of his father — a burden that soon gave way to the interminable misery of seasickness.

The quick sketches of his father in Konrad's popular books show a gentle old man in retirement: but this was later, and his father lived long. Both Konrad and Gretl remember him not so much for the challenge to his son — in which, in any case, he was unsuccessful — as for his zest for life: 'He was always happy, even when he was busy complaining about something!' An oil painting of Adolf Lorenz has long remained in his son's view. He took it to Seewiesen with him where it hung in the sitting-room of his flat above the laboratory; and in his own retirement it came back with him to Altenberg where it looks down upon his writing desk.

In his own way Konrad did submit to his father's will that he should become a doctor: he joined the anatomy department of Vienna University. This satisfied the medical requirement and with that Adolf had to be content. He was unhappy again when, much later, Konrad finally left the Anatomical Institute, and was only to be reassured when his son, then approaching forty, became a Professor of Psychology — that, after all, is a respectable branch of medicine.

Chapter 3

Jackdaw Spring and Danube Summers

It was a jackdaw that launched Lorenz on his career as a professional student of animal behaviour. It was, he says, sheer luck that he happened to hit so early on such an interesting species. On a chance visit to a pet shop while a student in Vienna, his eye fell on a dark bird in a dark cage. It was not a case of love at first sight; rather, he was moved by pity. He 'suddenly felt a longing to cram that great, yellow-framed throat with good food'. He intended to release it as soon as it was independent, which is what he did. But the bird stayed with him — and so, later, did its successors when he enlarged the colony, and with them a rapidly growing scientific reputation.

But first he had to fulfil the academic expectations of his father, at least to the extent of studying for a medical degree. Here too he was lucky, for his Professor at the Anatomical Institute was a man whose guidance set his feet firmly on the path of the scientific method that he was to apply to animal species of comparative anatomy. Professor Ferdinand Hochstetter was a teacher the youth could love and respect as a man and as his father in science. In contrast to the volatile enthusiasms and passion for new experience shown by Adolf Lorenz, here was a demonstration of the pleasure of patient discovery, set against a simple and unvarying routine. Looking back across half a tempestuous lifetime, Konrad Lorenz remembers the gentle anatomist with a deep and unalloyed affection.

The department chosen by Konrad, the student, was as close as he could get to the study of Darwinian evolution and still be in the Medical Faculty. He studied comparative anatomy with a will, and Hochstetter, at the time Director of the Institute, also taught him comparative phylogenetics — how to reconstruct genealogical trees from the similarity and dissimilarity of anatomical characters — and continued to advise him when he later sought to apply the same methods of comparison to the study of the characteristics of animal behaviour. Hochstetter came to regard his pupil as a pioneer of the application of his own methods in a new field. To Konrad, Hochstetter was a saint, a man totally dedicated to his science, the epitome of a happy man.

While Konrad was at the Anatomical Institute, Gretl entered his life in a second way. After nursing her mother, she had been able to conclude her high school studies, enter the University and was now studying to become a doctor. Gretl began dissecting in the same year that Hochstetter made his apt pupil a demonstrator; so Konrad was Gretl's instructor in anatomy. She, for her part, had already given some advice to him. In his last years at the Schottengymnasium Konrad had become an enthusiastic motor-cyclist, and had continued with it on his return from America in 1922. He now describes this cheerfully as 'one of the darker episodes of my life'. He rode reliability trials for the British Triumph Company until he crashed in the Semmering road race. He was unhurt, but when the practical Gretl told him firmly 'It's too stupid a way to die', he gave it up.

He had other friends to instruct him too, and his principal companion in most of his youthful enthusiasms was Bernhard Hellmann. Konrad's exact contemporary, Bernhard was born at the same hour in the same precinct of Vienna. He was also at school with him, and their friendship grew out of their both being naturalists. Bernhard shared deeply Konrad's passion for the hobby that Gretl had not joined in since those first ducklings they had shared when Konrad was six. Like Konrad, Bernhard was an addicted bird and animal keeper, fascinated by their behaviour. Together they puzzled over the question why the creatures they kept did some of the odd things they saw. When faced with something totally bizarre for which no explanation could be found, Bernhard would take an analogy from the world of motor cycling and ask, 'Is this how the constructor meant it to be?' They had seen how a motor cycle engine can 'spark over': a high voltage is built up and the electricity, denied its proper outlet, is discharged in the wrong place. In the same way, an animal could be doing something quite different from what 'the constructor' intended. But that thought was way ahead of its time, for even if the boys had known anything of the currently accepted theories of animal behaviour, they would have had difficulty in finding a place among them for Bernhard's idea.

Nobody could fail to remember one particular event. As a student in post-war Vienna Konrad had a study in his parents' town flat –

which meant as a matter of course that he also had animals there. As well as the inevitable aquaria there was a splendid capuchin monkey called Gloria. While Lorenz was in the flat she could roam freely within its walls; but when he went out she reluctantly returned to her cage. Once while the young man spent an evening in the town, Gloria at last managed to escape and proceeded to inquire into and thereby destroy a fair range of her master's possessions. The bookcase had been opened by use of its proper but minute key (an achievement that still excites Lorenz's admiration) and several textbooks removed for study. The top of an aquarium had been opened by the expedient of smashing a large bronze bedside lamp through its glass cover, thereby causing a short circuit and putting all the lights out. The pages of the textbooks were shredded and fed to the anemones in the tank. On his return to the darkened flat the student was greeted by Gloria's giggle from the curtain rod. Lorenz remarked on her strict attention to detail, the sheer physical effort involved for such a small animal; and also, almost as an afterthought, the expense.

Another event that occurred in his parents' flat might have been missed, or dismissed as unimportant, by anyone else, but to Konrad it was deeply puzzling. A hand-reared starling (which had therefore seen nothing of its fellow starlings' behaviour and, in fact, always took its food from a dish) sat on the head of a bronze statue, its head to one side as it seemed to examine the white expanse of ceiling above it. Then it flew up, snapped at something invisible to the young man, returned to its perch, beat the non-existent object to death, swallowed it, shook himself and settled down in peace. But there had been no insect; the observer, doubting the evidence of his own eyes, climbed first on a chair and then on a stepladder to examine the high ceiling at close quarters for insects he might have missed from below. Finally he had to accept that there were none now and, in all probability, there had been none before.

If birds gradually became the main object of his study this was probably because they are more immediately intelligible to man than are most mammals. Birds work largely with their eyes, the same sense organs that we also depend on most; eyes and then ears, the ears being used mainly for social contact. Apart from man, most other mammals think through their noses. Lorenz delights in a remark by Sir Julian Huxley that if we too were olfactory animals there would be no bird watchers, but in their place we would have mammal smelling societies.

Gradually, the young man's interest in birds sharpened further to concentrate more and more on Jock, his jackdaw. There was a hole in the wall of his bedroom at Altenberg through which his birds could pass to the larger attic beyond, and from there to the roof. Outside the window was a broad walkway between roof and parapet that Konrad could climb out to. Observations on the events in Jock's life began to fill his diary; his study became systematic and as time went by the body of notes grew. Meanwhile, Bernhard Hellmann came across a copy of *Die Vögel Mitteleuropas* by a distinguished German zoologist, Professor Oskar Heinroth. Here was a man who had looked at birds as the two boys did, for it described in detail the characteristic behaviour of each species. In Bernhard's firm opinion, Konrad's splendid jackdaw diary belonged beside work such as this. But

Konrad just went quietly on with his work—until one day the diary disappeared. . . .

The culprit was his fiancée, Gretl, who had conspired with Bernhard for the two of them to take matters into their own hands. They typed the notes out, composed a letter of humble submission above a signature that had been obtained by subterfuge, and sent the result to Heinroth in Berlin. 'And so his scientific career started', adds Frau Dr Lorenz. That was 1926; the paper was called *Beobachtungen an Dohlen* (Observations on jackdaws), and it was published early in 1927 in the *Journal of Ornithology*, Leipzig. It ends with the simple declaration, 'I intend to start a colony of tame jackdaws'. By his later standards the paper was very short—a mere eight-and-a-half pages of the journal.

In the year the jackdaw diary was published, Konrad married Gretl, the girl upon whom he had set his heart so many years before.

In due course and without undue exertion, he also received the degree of Doctor of Medicine—in 1928— although he made no move towards a career in that field. Instead, he remained at the Anatomy Institute where Hochstetter promoted him to Assistant, and concentrated his main efforts on the study of animals. He set the target of making a name for himself in three years. If he did not achieve this he would give up and become a doctor as his father wished.

Oskar Heinroth, as Bernhard and Gretl had anticipated, was impressed by the young man's observational powers; so Konrad now had a second academic mentor. He learned that the comparative methods commonly used in the study of animal forms had already been extended to behaviour by Heinroth himself: his 'breakthrough' paper had been written in 1910. Heinroth confirmed for him what Konrad already knew. The study of animal behaviour could be a science. Konrad had already decided to make it his particular business to apply comparative methods to the study of those behaviour patterns that were pre-determined by his animals' instincts. He had method, objective, observational skill, and the encouragement of one of the world's best minds in his chosen field. In the circumstances, it hardly mattered that he had never had any formal training in this new science of ethology before he himself had already become a practising ethologist. By the summer of 1928 the promised jackdaw colony was well established, and his studies of its social life occupied him deeply. Gradually their nest boxes took over the main attic and then were extended along the parapets around the house.

A picture of Lorenz at work and an idea of the understanding he achieved of these birds can be read either in the translated scientific account or in the relevant chapter in *King Solomon's Ring*, or—even better—in his eloquent marginal sketches adorning that book. These express his own delight at their skill in playing tricks with the wind. He shows the tolerance of jackdaws to his handling their featherless young, but also how his holding even a scrap of black cloth could have been mistaken by the colony for an attack on one of its members – the sight of 'something-black-hanging-down' would automatically trigger a mobbing attack and enmity for life. Lorenz had to be careful to avoid such a fate, and risked instead a local reputation for gross eccentricity by climbing out on to the roof of the house at Altenberg in party costume—as a fully rigged Satan with claws, horns and tail. So garbed in the first full disguise that came to hand, he could afford to be attacked by the rattling black birds as he handled the fledglings to fit their identifying rings, unrecognised by them as their ever-present friend. He noted the force of instinct in these attacks and also the role of learning as the older birds taught the young to recognise a potential enemy.

To the hand-reared birds the normally-clad Lorenz was an object of affection, and they would offer him choice tidbits of their favourite food. Among the birds themselves such behaviour is recognised as courtship, a prelude to mating. If he (not unreasonably) refused to take a wriggling worm in his mouth, it might be stuffed in some other convenient orifice, such as an ear. Living so close to the birds, he saw acted out the domestic drama in which a young, vivacious female intruded on the marriage of the leader of the colony, cautiously established herself at his side, and finally drove away the wife to replace her as queen.

One day that first colony was almost entirely lost, and he tells how the one remaining female perched on the weather vane and sang her heart out with 'come home' cries and so persuaded him to establish a new colony in the roof and eaves of the house. Lorenz is still half-amazed at his parents' acceptance of the mess and damage to the roof that these birds caused over the years. But by then their son had long and obviously been in love with 'the bird with the silvery eyes'.

His 1931 paper, 'Contributions to the Ethology of Social Corvids', he started while still a student and continued while an assistant at the Anatomical Institute. It was to be the first inclusion, many years later, in his collected papers—chosen because, he says modestly, it is 'quite a good illustration of how that kind of knowledge is gained, though it may require a bit of reading between the lines to gather how much time I spent staring at those jackdaws'. His pleasure at living close to the birds certainly does come across. The paper is all description and no theorising; that followed later. But his scientific reputation was established and Lorenz was on his way.

In contrast to all that he wrote on jackdaws, there is hardly a hint in his popular books (and little in his scientific papers) to show that he also kept herons. At one point, *On Aggression* promises a whole chapter on 'the loveless society of night herons', but the reference when it comes is brief and shows no particular involvement. The notes that he kept were never written up for publication, except for a single short paper comparing jackdaws and night heron societies prepared for an ornithological congress at Oxford in 1934. The courtship of herons is brief, and the resulting bond between male and female weak by comparison with that between jackdaws. The reason is simple: pairs of herons come together only to raise a single brood, and so have no need of the love that will survive a lifetime. All the papers from his three or four years' work on herons in the early nineteen thirties were lost in the chaotic conditions of railway transport across Germany in 1941. Lorenz calls this a tragedy—and indeed it was, for many of his most valuable scientific papers disappeared in that packing case. But it is also clear that he could never have loved the herons as he did the jackdaws before, or the geese later, or he would have written about them while he still had the notes. In fact there were two colonies including one of white egrets which never bred. His main group, the tame night herons, remained in the garden at Altenberg until he left during the war and there was no one to feed them.

While Konrad was still busy with his herons, Adolf Lorenz returned once again to New York and at the age of eighty wrote his autobiography. Dated the first of March 1936, he says in the foreword that it is 'a simple and human story of ups and downs'; and so it is. Couched in the easy, anecdotal style that also characterises his son's popular works, it is delightful, often funny and sometimes moving to read. The book throws a great deal of light on his own background and career, and thereby the background to his son's development both as man and scientist. At seventy Adolf had retired from his professorship in Vienna, and was not pleased to discover that an unpaid post had an unpaid pension—hardly adequate recognition for his years of service. But, says his son, his father's fortunes were perfectly well restored by his private post-war practice, and particularly from

America; although that was all lost again in the Wall Street Crash of 1929. However—and here we must return to Adolf's account—reward and recognition in the highest form for a scientist very nearly did come his way.

One day Adolf Lorenz received a request for all his papers on the 'bloodless' hip-joint operation. Nobel Prize deliberations are supposed to be secret, but a friend kept him informed on the progress of his nomination. His name went forward to the inner election and Adolf waited in elated expectation . . . until he heard that he had missed the Prize for Medicine by a single vote. 'What an old fool I was', he wrote, 'to have allowed hope artificially kindled to grow to a conquering flame!' His feelings are not left to the imagination of the reader. He flings himself into an aria of vindictive rage, heaping calumny on the head of that miserable miscreant, the one missing voter—but he pauses in mid-flight as though hearing the onslaught of his own fury, and turns sharply back to redirect it, almost as vigorously, against himself for such unworthy thoughts. He writes finally, 'You should not promise a man heaven, nor threaten him with hell if you are not quite certain about these places'. His name was put forward several times, to no avail, and in the end hope faded. On all this the son comments only that in truth this Prize can all too easily slip from the grasp, for among medical men there can exist the sin of envy.

In the whole of his autobiography, Adolf Lorenz has remarkably little to say about his dissident and rebellious second son, except for the circumstances surrounding his birth and a very few other brief notes—usually on matters where there was some effect on the father's life or actions that was worth mentioning. Such things, it would appear, were rare and slight. There is, however, little more about his dutiful son, Albert: a single sentence records Albert's second marriage and the birth of 'a lovely boy, Georg, who is very like his father'. He is proud that Konrad and Gretl, Albert and his own wife and assistant Emma, make a total of five medical doctors in the house: such a learned family!

In his autobiography, Adolf wrote that

Doctor Konrad Lorenz had made his study of the habits and psychology of the heron, which were kept in the garden free to go or come. Konrad, though a doctor of medicine, had preferred ornithology to medical practice. I was not over-enthusiastic about his choice and had deeply aroused my boy's anger when I said that it was of no great importance to know whether herons were more or less stupid than they were thought to be.

The assembled birdmen of the German Ornithological Society evidently thought differently and descended in strength on the heron-fancier's garden at Altenberg. Adolf tells us that he was thereby enabled to welcome their patron, His Majesty, Ferdinand, ex-King of Bulgaria. What Konrad thought of his father's cheerful address of welcome—to the effect that the best bird is that which is caponised, roasted and served at table—is not recorded, either by Adolf or his son. But Konrad was undoubtedly proud that his work should be recognised by such a visit—as much as Adolf was delighted that it should bring royalty to his doorstep.

There is one sentence in Adolf's account that is significant in view of the date it was written: 'Konrad contends that human psychology has much to learn from animal psychology and that there is no essential difference between these two branches of the science.' Was this originally conceived as Konrad's justification to his father?

The most productive years of Konrad Lorenz the scientist were those between the ages of twenty-three and thirty, and although important applications were to follow, his basic observations and discoveries were mostly written up and published by the age of thirty-five. 'I didn't discover much later on', he told me, 'all the important things were quite early.' This disarming avowal might be disputed by those looking at his post-war work, which would be a credit to any major scientific career. But it is also true that this mostly consisted of extensions of his earlier work to other animals, and a broader application of the principles and methods already established: in fact, the dotting of 'i's and crossing of 't's. In other fields, his major contributions were still to come, but for the basic science on which Lorenz's vast reputation is founded we must look to the twelve years from 1926.

His work as assistant to Professor Hochstetter came to an end when the old man retired from his Directorship. Lorenz did not get on well with the new man, so he moved into zoology. For a second doctorate he submitted a very long paper that he had already published, dealing with the mechanisms of bird flight and the different adaptations of wing forms. The title alone was three lines long, and the treatise was later republished in German as a book. At his oral examination he found himself facing a professor who, it soon became clear, had not read his paper but was strongly wedded to existing ideas. Konrad deftly switched direction, and gave the man the answers he expected. He saw no point in swimming against the tide of events, and in this manner gained his Ph.D. in zoology in 1933.

The young man also took an interest in psychology and at about the age of thirty his teacher, Karl Bühler, persuaded him to study the work of those who had written before Heinroth on the subject of instinct. Names like Spencer, Lloyd Morgan, MacDougall, Yerkes and Watson now appeared in his reading. These were all highly respected men, and some still are; but Lorenz was deeply disappointed by their ideas: to the young reader their authority was hollow. By his observational standards they simply did not know what they were talking about; quite obviously they had not studied animals with the insight that Lorenz had even before he became a pupil of Heinroth. None of them seemed to realise that behaviour could be studied in the same way as form, and they offered all sorts of ideas that Lorenz from his own direct experience could reject out of hand. The intolerance of youth, he reflected some years later, fails to appreciate that extreme and opposing statements, both of which by themselves may appear wrong, may when taken together contain some valuable truth: this had been said by the philosopher, Hegel. But even today, while quoting Hegel with approval, he is still not inclined to compromise. The way in which he rejected authoritative conclusions which did not fit the pattern of his own experience has always been an important part of Konrad Lorenz.

At that same time he was able to add one American name, H. S. Jennings, to the short list that received his young approval. Jennings had observed that in each separate species there was a characteristic 'system of actions'—which is what Lorenz also saw. Even so, in his early papers Konrad made valiant efforts to fit his observations into existing theories of animal behaviour. In particular, it was some years before he was able to clear away from his mind the debris of earlier ideas that had explained behaviour patterns as automatic reflexes.

A reflex, at its simplest, is illustrated by the jerk of the leg that follows a tap at a particular point below the kneecap. From the start it was evident that many patterns of behaviour were more complicated than that; so the 'chain reflex' was devised. In this a stimulus from the environment triggers part one of the behavioural response; this itself triggers part two, and so on until the whole pattern is complete. Many actions, like walking or swallowing, can be explained well by such reflexive sequences, but physiologists of the time seemed to believe that all behaviour must operate at this superficial level, like a slot machine that always produces the same, unvarying response. Lorenz was slowly moving towards an understanding that much behaviour comes from some deeper source. He saw cases (like that of the starling that he had kept in his parents' flat) where it seemed to arise spontaneously, with no stimulus at all: it was there just waiting to happen. If the stimulus that would normally trigger the action did not come at the appropriate time the action would be delayed, but it could still occur without any external cue. In other words, he concluded, an instinct for the action exists, and eventually in one way or another it is likely to express itself.

The older Lorenz belabours himself for not understanding earlier what most people in his position might never have seen at all. If he was slow to formulate the new idea—and perhaps only Lorenz would be so impatient with himself as to suggest slowness—it would certainly be reasonable for him to accept for a while such ideas as came to him through (if not directly from) the students of behaviour he did by now respect—Oskar Heinroth and a second American, Charles Otis Whitman. It was not until ten years after Lorenz met Heinroth that they both heard of the work of Whitman who had pronounced in Chicago in 1898 that comparisons between species was the proper way of studying animal behaviour.

In 1919, Whitman used a behavioural characteristic to define membership of the order of pigeons: without exception, these birds use peristalsis—a progression of muscular contractions—to pump water upward along the oesophagus when drinking. Lorenz says that he learned much from a correspondence over several years with Whitman's pupil, Wallace Craig, who described in 1918 the way in which instincts express themselves as appetites and aversions.

Many years later, when Lorenz came to select some of his early papers for republication (and also for translation into English), he drew his readers' attention to the fact that, in each case, the conclusion was weaker than it could have been, but that the end he should have reached is embodied or almost implicit in the start of his next paper of importance, written a year or so later; or, if not there, in the major theoretical papers on the nature of instinct that were to follow.

The work that he was doing—in his spare time at first—was deceptively amateurish in appearance. In fact, Lorenz himself would delightedly accept the label 'amateur', for it is the true amateur, the lover of his subject, who is most likely to become the real expert in it. It is also true that in his lack of early formal training he was an amateur in a second sense of the word, although he now set out resolutely to fill the gaps in his knowledge.

But his life was far from being all work and no play. In both work and play he used his riverside surroundings to the full and enjoyed especially the mastery of one particular element of his environment — the Danube. He regularly swam across its swift but steady currents, which present no great problem to an accomplished swimmer who knows their tricks. Indeed, every summer Sunday many trippers took the train from Vienna to Altenberg for the pleasure of diving into the yellow waters. Crossing and recrossing the river was good exercise — the slower the swimmer the longer his walk along one bank or the other to get back to his starting point.

The second component of his mastery was achieved in a way that his friends quote as typical of his vigorous, even aggressive, attitude towards the challenges offered by the world about him: he also studied for and obtained the 'patent' of a Danube riverboat pilot. It was, he says, the toughest examination he ever sat (which may only mean that no other examination cost him much effort). The explanation for this surprising act is simply that at the time it was necessary. It was in the days before there was any provision for private pleasure-cruising on the Danube. A forty-feet longboat that had once served as an Austrian Man-of-War was 'going for a song' as military surplus equipment. Konrad and Gretl bought it and with the help of a local boat builder converted it into a cabin cruiser; a cheap second-hand Mercedes tractor motor provided the power. Konrad buckled down to his studies and qualified to pilot a steamboat of up to twenty metres and two thousand horsepower — the smallest craft for which a patent existed. Then they cruised to Budapest and back on holiday.

In the thirties Lorenz had no animal institute or field station other than the family summer home at Altenberg. When they talk about that time, Konrad dutifully acknowledges his wife's tolerance — which followed that of his parents but with a more practical justification. Her husband's professional reputation was already secure in her own mind, if not in his father's. To Konrad's husbandly praise she replies that for most of the time when the animals were present in their greatest numbers she was out pursuing her own career as a gynaecologist. Their two older children, Thomas and Agnes, had been born in their first two years of marriage, then she had qualified as a doctor (four years after Konrad) and immediately began earning money at a time when her husband had no secure income. For the rest, she and Konrad's mother put up with the neat circular holes chewed in sheets by rats collecting nest materials, and put cages around the children when they had to be protected from the inquisitiveness of monkeys.

The method they devised for looking after the young children, the 'reverse cage' principle, was designed to give the more active and energetic animals as much freedom as possible; the two infant humans could manage well enough with a degree of confinement. They needed less protection from the dogs, for these have a built-in capacity for recognising the young and a strong inhibition against harming them. As Lorenz records in *Man Meets Dog*, even the tiny adult chihuahua can recognise the 'babyhood' of a young St Bernard many times its own size, and any grown domestic dog should be able to distinguish a human baby and react to it as it would to a puppy. On one occasion, six-year-old Agnes marched boldly into a dogfight to separate the participants, was knocked over in the action a few times, but emerged otherwise unscathed and still without fear; only a little puzzled that her authority was less easy to establish than usual.

Thomas and then Agnes, like their father, found Altenberg a good place to grow up. But unlike his father, Thomas did not become 'imprinted' on animals: he found them amusing and always the subject of interesting conversation but, given the choice, he preferred to play with the train set that his father had bought him. Konrad played with his son and the trains too; and when Thomas left home after the war and took these with him, his father got himself (for the grandchildren, of course) an even grander layout, using the new miniature gauge. A three-dimensional structure with green hills and tunnels, it stands in his study still.

Thomas's recollections of childhood in a happy household include the awed respect with which everybody treated the 'Herr Hofrath': this was the aging Adolf, under the title (Counsellor to the Emperor) that had been granted to him in the previous century (an event that, needless to say, is fully recorded in his autobiography). Although by his own nature he was readily approachable by the children, they also observed the speed with which his every whim was fulfilled, as befits an old but still great man. Thomas also heard talk of the lack of money; and later the hope that the autobiography—in English for America, where it could hardly fail to sell—would restore the family fortunes. It didn't.

Among Thomas's earliest memories is one that caused laughter at that time: the antics of the young hotheads who illuminated the hillsides around Vienna with crooked crosses of blazing spruce . . . and their leader, that ridiculous fellow with the funny moustache whom everyone dismissed as a comic turn.

From the interests he shared with his father, Thomas learned not of imprinting and instincts but of torque and revolutions, and so grew up to be a physicist, though later with a half shift back to biophysics.

For all his emphasis on instinct, Konrad Lorenz accepts and indeed emphasises the role of experience in moulding the young of any species. The early influence of his own father and also his father-in-law was strong and later he formed an affection for his teachers Hochstetter and Heinroth. In the influence she wields, his wife occupies a remarkable place in his life, and the roots of this lie not only in the bond of marriage but also deep in his childhood. He says, 'My wife and I were more or less siblings'. (An ethologist might note that it is highly unusual for one animal to marry another that when young has shared its nest. But the range of possible and acceptable human behaviour is vastly wider than that of any animal.)

The word 'marry' may have as much relevance in the world of jackdaws or geese as it does in human society. Among his favoured species of birds, pairing is for life: compare this with the human institution of marriage and you might conclude that the birds actually have a greater capacity for monogamy than is displayed in many human cultures. In his goose work Lorenz has shown that a gander will mate only with a suitably inspired female; and that in turn the females would accept only adequately dominating males. This suggests 'a human analogy that no women's liberationist would particularly welcome' an American interviewer noted drily.

But 'There are many roles you must play', Gretl says, and these include '. . . to defend some things against the husband'. This could also mean 'to defend the husband from unseen attack', which in turn requires an understanding of what he does. It is also her job to criticise his work—'Yes, yes, even that!' her husband comments cheerfully.

Does this underrate or oversimplify the influence that she has on him and its effect on his life? Some of their friends would say that it does. They say that he has been in many ways fortunate in the stability that she gives him: she has a dampening effect on his wilder oscillations. Like his father before him, Lorenz has the 'crazy' ideas—this is one of his greatest strengths— and Gretl brings him down to earth. He is wild enough in his enthusiasms of intellect or action that some restraint must often be good. But for associates who would sometimes like to knock a crazy idea around for a little longer to see what else it might lead to, it can be frustrating to find that Lorenz himself has discarded it overnight. This may often be after rational discussion of it with his wife, who is more conservative than he. But overall, and generally, the combination is more powerful than Lorenz on his own. He consults her on the style of what he writes and has to say in public; and in the shakedown that follows, the surviving ideas become both reasonable and communicable. While Lorenz often treats his wife with affectionate casualness, he certainly finds in her a touchstone of common sense. More than this, she has been his 'Glücksbringer', his bringer of luck.

Another strong element in a child's early experience is the place where he was brought up. If, as with Lorenz, it was a very happy childhood, and the child had the mental and physical equipment as well as the inclination to interact with that environment to the full, we might expect the sort of nostalgic reaction that Lorenz still expresses towards Altenberg. He contends that he sees in people one remarkable effect of irreversible 'imprinting', expressed in their preference for a certain type of landscape: what you lived in during childhood you want forever. 'This, you see, *is* imprinting', he insists.

Chapter 4

The Goose Years

More than any other of his animals, it is the greylag goose that is most closely identified with Konrad Lorenz. This reputation rests largely on the pictures of him among the geese at Seewiesen that went with post-war interviews, which were stimulated in turn by the public interest aroused by *On Aggression,* and by the chord of comparison between goose and human societies that resonated so readily in the copy of the interviewers and the minds of the readers. But all that came later, and with geese other than those he had originally studied.

Lorenz himself will tell you at once that 1935 to 1938 were his goose years, the real goose years at Altenberg where he worked with broods of goslings long since dispersed by war. Those were also the years of his most powerful scientific papers, rooted in the work he had done before but confirmed by his observation of geese. This whole period he calls his 'goose summer'; it was a woodland, village, grass and river summer. Sunshine plays on his memories of that time, and fresh, clean country air flows through his descriptions.

He is relaxed and confident, and the nature of his work is in harmony with the gentle pace of life around him. When he can, he pours out his ideas on paper, but he also has plenty to do in caring for and watching his goslings, and with this he also has time to think. He has described himself both then and since as a 'peasant scientist'.

The peasant is a small-scale farmer, understanding but also managing each significant detail of his biological environment. For the true peasant the significant parts of that environment are those which affect the production of food or other saleable materials; for Lorenz, it was the production of knowledge about the behaviour systems of animals. But the attitude and approach of both are similar: each walks in his tiny kingdom, ruling it gently but firmly, directing it to work for his ends. This is different from post-war farming which has achieved more—in the short term at least—by applying stronger controls, and propping up with chemicals the unstable system that results. The laboratory study of animals is a little like that: a narrow range of results comes quickly and in repeatable experimental conditions. But there are disadvantages. One is that the probing of systems in stress can produce different results from their observation in near-normal conditions. To discover the ways of some new animal, or the detail of a behaviour pattern, it is better to start with what happens in an ordinary course of events, if only to have a point of reference.

Even so, Lorenz was not so naive as to suppose that he, an outsider, was not interacting with his animals. Indeed, his most commonly chosen method of study, imprinting, was designed to bring the animals to him and for them to accept him as one of themselves. But there could be limitations to this approach too. For example, suppose that the animals were not simply accepting the human as one of them, but were instead modelling themselves on him? Lorenz was fortunate in that his particular line of study should not be too affected by this possible inversion of effect: he was concentrating mainly on the study of those aspects of behaviour which might be inborn to the animal itself, and if in this artificial situation they still emerged, then the behaviour was all the more likely to be inborn.

Each day as he walked, sat, canoed or swam he could be seen followed or surrounded by a bizarre and apparently random selection of the various creatures that he had chosen to study; and when he returned home he would walk among others, those which less easily formed attachments, or had not formed them to him. However his interest in an animal may have arisen in the first place—and this may in part have been by the interplay of chance and curiosity—his chosen subjects did in fact form a coherent and rational array. The different species fell into several groups. First, there were those which were in their own right the central objects of his study; initially the jackdaws, then the herons, and now the geese. Second, were the closely related species: ravens for comparison with jackdaws, or mallard ducks to watch out of the corner of an eye while looking at geese. These showed not only what the ducks had in common with their geese cousins but also what they had developed differently: he could ask himself 'why?' Heron society was markedly different from that of jackdaws or geese; again 'why?' Then there were the species unrelated to jackdaws or geese, but which had similar elements of behaviour. This allowed him to look for patterns of behaviour to which evolution came independently in different species: such features of behaviour are said to be 'convergent'.

The goose work actually began in 1934, the year that Alfred Seitz, among his earliest students, first came to Altenberg. Seitz was older and perhaps better motivated than most other students, for after leaving high school he had spent eight years in unsatisfying jobs, but in all the spare time he could find he explored the margins of the Danube for its bird life. A good amateur photographer, he gave up his job and took his bulky, old-fashioned camera to the shallow, reedy waters of the Neusiedlersee, south-east of Vienna, where he could seek out and photograph herons, glossy ibis, spoonbills and their nests, and the wading stilts and avocets feeding on the rich harvest of crustaceans in the water around their legs. But when he began to collect material on their natural history for magazine articles he soon discovered he needed a better understanding of their behaviour. It was to help him evaluate his observations on the herons that a Viennese ornithologist suggested he pay a visit to Lorenz at Altenberg. Seitz recalls, 'Even that first encounter with him was decisive for my subsequent training and later career' — a statement echoed by many since.

Seitz attended Lorenz's lecture course, 'An Introduction to Ethology', worked at the aquarium on his own studies of the behaviour of cichlid fish (the African mouth-breeding fish that later set Lorenz off on his own studies of aggression). And for the start of Lorenz's goose studies it was Seitz who obtained the fertilised eggs.

Lorenz's method required birds of the true wild species that he would raise from incubated eggs and tame for himself. Domestic animals that had degenerated genetically from their original state would be useless, and if 'greylag' eggs were bought in a shop there would be no way of telling whether the stock was truly and completely free of the behavioural genes of some barnyard waddler from a few generations back. Seitz knew where to find exactly what Lorenz wanted, for he had seen the breeding places of the shy, wild geese that in those days lived deep within the Neusiedlersee swamps, far away from the likelihood of human disturbance; and he also knew the approximate time of hatching. Having found and taken a clutch of eggs, Seitz then had to get them back, still warm, the sixty miles to Altenberg.

To their joy, the incubated eggs hatched satisfactorily and were imprinted on Lorenz as their parent: these were the geese featured in Seitz's film shot in 1935. One, called Martina, was given special treatment: Lorenz kept her with him longer than the others in order to study imprinting in more detail, accepting with this the merciless restrictions on his own life that such unending attention demanded. Once, as Lorenz was at his typewriter, Seitz saw him mutter an 'ouch' and reach inside his shirt. Martina, keeping warm there, had woken and begun to pluck the 'grass' on his chest—thereby providing for the student a direct observation of the power of instinct.

The initial success demanded an expansion of the colony, so during the next Spring Seitz set to work again. This time he decided to approach nesting grounds on the Hungarian side of the lake. The first part of his plan was to 'lure' Lorenz into lecturing to the ornithologists of Ödenburg on the Hungarian bank of the Neusiedlersee. Next morning, their friends led the party into the reeds of Lake Fertö, where the unsuspecting boatmaster took them through a long, narrow channel through the reeds, and straight to a greylag nest. Making as little disturbance as possible, each of the group in turn went to peep inside. Seitz carefully arranged that he would be the last to go—and while he was there, 'in the interests of science' he removed the already well-incubated eggs ('the poached eggs', quipped Lorenz, enjoying the pun that appears in translation). No one suspected the poacher but Lorenz, who noticed the solicitous care with which Seitz now treated his rucksack, and the culprit confirmed his leader's suspicions by handing over all the 'poached eggs' that somehow had to be got back home.

The Altenberg goose stock soon comprised one wild gander, some ten goslings that were raised by a domestic goose in the garden, and ten that were mothered by Lorenz himself. His own brood trailed him around the house and garden and on (for them) long and strenuous walks down the road to and along the Danube. He expected that the goslings that stayed in the garden with their flightless foster-mother would prove the better stayers and that his own wandering flock would eventually be lost. But as they grew up, he discovered the reverse to be more likely: if a goose is taken a great distance—say a hundred miles or more—and then released, it can find its way back only to the general area of its home, and must then rely on familiar landmarks, of which the stay-at-homes learned none.

To the geese that were not imprinted on him and so did not follow him he gave numbers; those that were and did had names. For over thirty years, with these and his post-war geese (even when their care and study had largely passed to others) he was to have the pleasure of calling them in their own language and seeing their flight towards him. They derived confidence from his presence in unfamiliar territory, but apart from this special relationship much of their behaviour remained close to that of the untouched wild bird.

For Lorenz, his choice of geese for this new study reflected in part the sheer approachability of the species. Geese are easily socially imprinted on man without becoming sexually perverted. In this they were far more satisfactory than jackdaws, for 'if you raise jackdaws as tame as our geese are, they all court humans'. Beyond a certain point, this redirection of their sexual drive makes research on their social behaviour impossible. A third reason for looking at geese is that what we learn from their behaviour may be more relevant to man than is the behaviour of some of our closer relations in the animal kingdom. It happens that the primates, which might be expected to have greater similarities to man, form different types of societies with different personal relationships from ours. Assuming that man is not exclusively the creature of his conditioned responses — and Lorenz is the leading exponent of the view that man is not — we can learn a great deal from the analogies that may be drawn on the basis of similarities between the behaviour of man and animals.

Geese are also relatively simple to study — always an advantage in research — and quite remarkably close to man (as Lorenz sees it) in personal friendship, in loyalty and enmity towards friends and strangers, in relationships between parents and young, and even in married love. Lorenz once watched the behaviour of his greylag Ada when she lost her mate. He saw her flying in ever widening patterns of search, crying her long distance call until, eventually, all hope was gone. The bird now behaved as though deprived of all purpose in life. She lost courage and was fearful in contact with other geese. Her face altered, the eyes deepening in their sockets and the muscles around them sagging: the stricken bird knew grief. Eventually, after repeated widowings, Ada became promiscuous. One visiting ornithologist looked into her face and said without prompting, 'That goose must have been through a lot'.

Is this way of talking of his birds unduly anthropomorphic? If so, Lorenz qualified it with an adage that he quotes: 'There's an old Chinese saying, "There is all animal in man, but not all man in animal".' Anthropomorphic words must be used with caution; but they are often, Lorenz contends, the best descriptions that can be found. At the very least, their use makes it quite clear what analogies are being studied. More likely our 'human feelings' are our own experience of a system common to animal and man that serves as a prompt to behaviour.

Lorenz notes that following Darwin, his own respected teacher Oskar Heinroth attributed instincts to man—instincts deriving from his animal ancestors. In 1910, when Heinroth had been studying the duck and goose family, he drew particular attention to the striking similarities between bird and man in their social behaviour. The analogy was itself an argument for the existence of instinct in man. To argue this in reverse: Ada the promiscuous goose had become what in humans we would call neurotic. Her instinctive patterns of action had served her well when her life was proceeding according to the plan to which her instincts were adapted. At that time, they meshed neatly with the corresponding actions of all the geese she met in their common environment. But the complex interdependence of her own linked patterns had been disrupted and she had become a misfit in her own society. Ethology is full of such examples, as is human psychological medicine. Here is a second argument for man and animal having similar mechanisms at the roots of their behaviour.

Greylag geese may be like man in another respect, too: the length of their life. Lorenz is no direct authority on this, as the geese he raised at Altenberg were with him for only a few years, so that even in 1950 he was quoting twenty-five as old for a goose—although he noted with surprise that, when eaten, the twenty-five-year-old casualty of repeated and finally incautious dog-baiting tasted remarkably tender and young. He now observes that no reliable upper limit can be set for a greylag's life, since no goose-keeping institution has yet outlived its geese. A goose is mature at two and appears no older twenty years later; and records of geese sixty or more years old have been published in Russia and Canada. A more common animal for biological study is the mouse: its entire life-cycle can be followed in two years, so experiments in genetics involving several generations are possible within a reasonable time. A bird of comparable size (a canary, for example) lives twenty times as long as the mouse. For Lorenz's work he must accept the disadvantages as well as the advantages of his chosen bird's very much longer lifetime.

To raise his own geese at Altenberg Lorenz had to fit his own life and studies to their daily cycle of activities. Geese, like ducks, have no prolonged periods of sleep. They rest in the dark hours and again for an hour or two after their regular midday bath. Even then, unlike many other birds, they are not deeply asleep but are aware of even the quietest approach. For a number of years before and after the war, the day's work began with the early light, when Lorenz would call the birds to him, work with them, and share their morning swim. His duties continued, on and off, until his charges settled down for the night. When, during his pre-war goose years, he resumed his lectures at the University, the two activities could be fitted together only with the greatest difficulty. Seitz noted how, knowing the routine of his charges' day so well, Lorenz could show visitors round and explain not only what was going on at the time, but also what the birds would be doing during the next few minutes. He could then enjoy each guest's astonishment as the predicted behaviour started.

Lorenz watched over a cycle of behaviour that includes courtship (and 'marriage'), mating, nesting, hatching and then rearing the gosling family. He followed the growth of the young and saw that even before they are sexually mature they begin to look for the life-long partner with which they will repeat the cycle. In the Spring, the visitor to Altenberg (or to one of Lorenz's later Institutes) might be shown a young gander swimming and chasing after an unattached female. The youngster will sometimes catch up with the object of his interest and chatter quietly to her; if he is in luck, from time to time she will reply. During the several weeks of this early courtship he will also look for opportunities to dash off and attack other young ganders, for a good scrap with victory as the outcome will demonstrate his boldness and strength. After acquitting himself with honour the young gander will return, calling triumphantly, and his intended bride will join him in his excited chattering and the highly significant 'triumph ceremony'. They dip their necks, lowering and pointing their heads almost like a signal of intention to attack, but direct this threat gesture past each other and add the same chattered greeting. To create the strong bond of goose marriage, the courtship of these young greylag geese is long, and their attachment will be reaffirmed by the ritual of greeting which therefore has more than just an immediate value as a signal.

Lorenz talks of 'love', 'betrothal' and 'marriage' in his geese in almost human terms—quite deliberately, as we have seen. Occasions of love at first sight (with the appropriate signals of 'betrothal') are impressive enough, but characteristic also of his birds (jackdaws as well as geese) is the way in which the demonstration of a previously unrecognised love may follow the reuniting of two close friends after a period of separation. The analogy with human experience—his own also, as he recalls—is striking.

When goose courtship at last moves on to mating, both partners dip head and neck in the water, a clearly understood signal that may also serve to strengthen and synchronise their inclination to mate. As their necks are dipping almost together, the female suddenly flattens herself on the water and the male mounts her.

She lays an egg a day, perhaps for as long as a week, and it was Lorenz's job (or, in more recent years, his assistants') to check and remove any infertile egg that might either cause the goose to go on sitting when the parents should be up and away with the hatchlings, or eventually get broken and draw predators. Lorenz holds each egg in turn against a tube (formed by rolling up a magazine), through which he peers to see the growing foetus against the light. The goose sits alone on the large white eggs with little time to get away and feed, while the gander stays nearby to defend the nest against attack or intrusion. After nearly a month of incubation the eggs begin to 'talk' back to their mother, and a human who cries 've-ve-vee, ve-ve-vee' to the egg will also get a trilled chirp in reply. The eggs hatch almost all together, and within a matter of hours, each bedraggled hatchling emerges as a dark, damp, sticky ball of fluff already squawking in a frequently reiterated conversation involving both parents and all of their offspring, whose very survival depends on quickly learning to recognise and stay close to their parents. This pair will protect them; other geese will drive them off.

About a day after hatching they can walk. They stumble and chase after their mother as they see her move away. When the parents stop, the hatchlings turn to nibble at grass or almost anything else that could possibly be food. The mother can be relied on to lead them to food, for she too is hungry. If two foraging families get too close together, the parents will threaten and drive the others off until there is enough separation to prevent the young from getting mixed up. Soon they are on the lake, the mother leading the hatchlings which follow her, and the gander watching over them from behind. The mother bathes herself, and the chicks follow suit: they dive beneath the water, rehearsing a manoeuvre that will stand them in good stead if they are attacked upon it.

Among these geese there are firmly established signals for danger, threat, greeting, the preparations for mating, and so on. Each of these involves the use of the head and neck in some characteristic manner: a raised neck, for example, steady and attentive, signals danger. A phalanx of geese — a mob — that has moved closer together for protection against a common danger (which may be a fox or a dog that they do not trust) is an impressive sight.

I saw this mob reaction against a coati (a small raccoon-like creature) being taken for a walk on a lead. The geese gathered at the first sign of danger and kept about ten feet away from the coati, a potential predator on their eggs or hatchlings. Their necks were raised in unison, marking the danger, and when the coati passed in front the mob heeled with it. The animal was terrified, and without the persuasion of the lead would not have gone anywhere near the geese. There was no actual attack, but a mob of geese can achieve its aim of intimidation without resort to violence. In human society a picket line can have the same effect on any marked individual.

At Altenberg there were many animals living in close proximity, and the geese showed that they could soon discover which was their potential enemy and which harmless. Lorenz, by watching a goose react to a dog, could discover whether the dog was considered reliable or not. On one occasion, when deep snow at Altenberg limited the choice of paths through the garden, he saw three running, barking dogs leaping over a line of geese, which stretched their necks and hissed — on principle, as it were — but clearly did not find it worth the effort of standing up to make their protest.

Geese show their dislike of anything that offends against the rules of their society, and any abnormal behaviour is met by strong pressure to conform. A mother without an attendant gander is a social outcast, and all outsiders are suspect. Their reaction against aliens is most obvious when strange geese come into an area that already has a flock in residence. The strangers will be driven away by clearly understood signals that are mainly visual. A gander will stretch his neck low over the ground, advancing or running at an enemy. A group of invaders may drive back stragglers that have separated from the home group, but the resident geese will consolidate ranks to confront the enemy. Having more to lose, the defenders are the likely winners despite numerical odds against them. It is very rare that any serious, physical combat occurs in such an encounter; in a ballet of advance and retreat the birds are well able to assess the strength and fighting intention of their adversaries. After each individual skirmish, the self-declaring winner returns to his mate proclaiming his victorious intent, again with the ceremony of displaying the outstretched threatening neck as he runs past his companion. She responds in kind, inciting and urging him to greater feats.

Attack and defence may occur between different kinds of geese and swans, since these species are sufficiently closely related for each to recognise the others' threat to their livelihood. Similarly, bonds of friendship can form across the boundaries between the half-related birds. But while swans invading the territory of geese would drive them out, neither swans nor geese pay any attention to ducks unless they actively get in their way. The two groups may occupy the same broad living space. To geese, ducks are simply 'not people'.

Within the flock itself there are strong and weak, brave and timid individuals. It is a stable society where each bird knows his place, and for this it follows that it must recognise and develop a personal relationship with each other individual, even in a flock of several hundred. This in turn produces a rank order—sometimes called a pecking order, though this term could be misleading in a society where it is rarely dependent on actual pecking or fighting. A goose demonstrates its superiority to another by its threatening behaviour, and the hatchlings in turn learn from their parents when they may attack and when not. A parent from the superior family does not instigate an attack on the goslings of the inferior group, but he ritually defends if any member of his own family is attacked. This takes the form of a simple demonstration: the superior goose simply bends forward and gently plucks at the feathers of the attacker from the other family.

In a goose colony there are three independent hierarchical systems. The first depends on a mutual assessment of the relative individual strengths of two ganders: on this basis, when the two meet alone one will give way before the other without dispute. But in a family altercation where the nominally weaker gander had achieved a greater breeding success with, say, six tough sons ranged against a couple of runtling daughters in the other family, the situation can be reversed, and the opposite result achieved. Within the family there is a further rank order, settled by the outcome of squabbles in the early days of the hatchlings' lives. Here again it rarely requires anything so strong as a peck to reinforce the order, but it is constantly visible in the manner of family greeting ceremonies. The threatening element in the superior's greeting is more marked and he directs it closer to the other more obviously submissive bird. But this order does not carry over to the outside world, where a junior within the family can display high courage and so gain disproportionate prestige. He may be aided in this by an initial defensive posture that serves better in real fighting; and by the fact that his opponent is used to receiving submission and thus has little experience of having to make his threats good. The family junior knows nothing of such easy superiority and may fight more tenaciously and effectively. . . .

And so the description goes on: this is the barest outline sketch-map of the world of his geese that Lorenz has observed. Some of the social behaviour could only be studied later, with the much larger numbers of geese and ducks at Seewiesen, but in his early years at Altenberg Lorenz discovered much of its richness and ritual in the lives of his less than twenty geese, and in his thriving colony of jackdaws.

His aim was not purely descriptive: he was looking for the structure of behaviour, to find what scientific laws may determine its emergence in the developing animal and govern its form in the mature society. His discoveries were not quite like the simple laws of the physicist, but they differ in detail rather than nature: animal behaviour has proved more complex and variable than the world of physical measurements. So it transpires that many statements about behaviour have subsequently had to be modified to take account of the exceptions, and terms in common currency have changed their meaning, much to everyone's confusion. Before examining the ideas themselves, let us look briefly at the problem of the words that have been used to express them.

Some words have been so muddied by association with outmoded explanations that they have since become unfashionable. As with the drifting succession of euphemisms for the 'lavatory', 'WC', 'toilet' or 'loo' in Britain, or 'the john', 'restroom' or 'bathroom' in America, the word 'instinct' has been superseded. But in this case the succession of terms has also been accompanied by gradually altered shades of meaning. Complex technical language has been devised to get round the problem, but it makes the subject unintelligible to the laymen and until recent years has separated the practitioners of different disciplines (particularly the ethologists and psychologists of various schools) who really do need a common language. Even when common terms were agreed for subsequent use it remained difficult for either side to look back objectively into a past cluttered with rejected terminology. The terms 'drive', 'propensity', and 'instinct', and others like 'inborn', the later and apparently more acceptable 'innate' or (as Tinbergen prefers) 'environment-resistant', belong to a group which have so much overlap of meaning that the layman might reasonably be amazed by the amount of infighting that has attended the use of one word or another over the years. For example, 'propensity' may be acceptable to some who would shudder at talk of an animal's 'drives', a word that Lorenz inherited from Heinroth.

Arguments around the word 'instinct' reflect in part a debate about whether the word describes something simple, coherent and concrete in the way that the words 'limb' or 'organ' refer to parts of the body that are recognisably whole in themselves (like arms, legs, lungs and eyes) or whether it is merely a temporary but convenient concept, a way of handling an idea (like 'ego' or 'id'), something which would have to be redefined if we tried to look for its physical basis, or whether again it might be something that lies between the two extremes. In this book, despite the reservations of younger ethologists—and perhaps even my own—it will be simplest to use the word 'instinct' as Lorenz did originally (and as he really still prefers) to mean something actual and physical, a given and unchanging part of the animal.

The animal's learning, its experience in every form of interaction with the outside world, also contributes to the overall complex of behaviour of the mature animal. In his studies of geese and jackdaws Lorenz believed that he could see elements of both instinctive and learned behaviour, although he concentrated more of his interest on the former. He saw that many of the actions of his animals could be explained by propensities for those actions that must already have been there at birth; that is, they were innate or inborn. These instincts hidden somewhere within the newborn animal would have to be genetically determined in the same way as the form of the animal.

In talking about such matters Lorenz may say, 'We are particularly interested in the inborn—the phylogenetically programmed, that's the modern word for it—behaviour patterns in animals, particularly those of underlying social structure'. The phylogenetic programme (the pool of genetic information that has been fashioned by the evolution of a species) is available with minor variations to each individual from conception onward. In modern practice this can therefore be simplified to equate 'innate' with 'genetically determined'. This has the practical advantage that you can start thinking about sensible experiments which link behaviour with genetics. And indeed, behavioural genetics (sometimes called psychogenetics) is a flourishing young science related to the work of Mendel and Thomas Hunt Morgan as Lorenzian ethology is to the ideas of Charles Darwin. Today we can half visualise how determinants of behaviour might be written into the genetic code: we imagine DNA spirals telegraphing and projecting their messages forward into something that, given the proper external conditions, expresses itself eventually as action. It would be a complex process, involving the setting up of structures in the brain and the central nervous system, the availability of chemical messengers such as hormones, and—often—rehearsals of the capacity for a response. The interaction of genes with environment can programme the brain—quite literally making up its mind—in preparation for the behaviour pattern, or a possible range of behaviour patterns (including 'changes of mind') that will follow.

The thirty-year-old Lorenz's ideas about all this were necessarily simpler but developing rapidly. For us to see it as he did we must look back as though through a time machine. There were as yet no codes of amino-acid bases strung at various points on long molecules of DNA to aid visualisation and make it concrete; you had rather to think of separate, individual and coherent calls to action, each wrapped up and located in its own special box. What eventually emerged from the boxes hidden within each individual would be subject to the recognised laws of mutation and natural selection. This could change the shape of any of the genetic chunks, and hence the message, without affecting its essential unity: behaviour could evolve.

If we were to open one of Lorenz's boxes out would pop some more or less fixed pattern of action. Each animal would be born with a number of these boxes of behaviour which would qualify it as a member of its species as clearly and firmly as does any physical structure of its body. At the simplest level: unless the right mating behaviour is followed the animals may not mate. Such a coherent building block of behaviour has been called 'inherited co-ordination' (in German), 'fixed action pattern' or 'fixed motor pattern' (in English) and 'drive-operated activity', 'instinct activity' or 'innate behaviour pattern' (in Lorenz's early papers).

Whatever the name, as Lorenz at that time saw it this behavioural element would at first remain latent as though secured in its box with a combination lock. He argued that the correct combination for its release would have to be provided by an event or series of events which (from the previous experience of the species) could be expected to occur in each individual's environment. With the final click of the combination lock the behaviour emerges in an action that will generally promote the survival of the individual and, since this will generally be a common characteristic, of the species, too. Even before that happens, earlier parts of the combination may cause the animal to seek the precipitating stimulus if it does not appear in good time.

This search (sometimes called 'appetitive behaviour') appears purposeful; the animal behaves as though it knows what it is looking for. The manner of the search depends on its current environment, but the goal remains the same. In comparison, the action that is eventually released has a mechanical rigidity that defies change: the animal performs it with the lack of awareness of a sleep-walker.

The stimulus may itself come from the animal's physical surroundings: it may be as simple (or complex) as the arrival of Spring. Or (and here was an important new concept) it may come as a signal from another animal of the same species, from an organ — a 'releaser' — which is adapted specifically and precisely to perform this function and so to promote survival.

Lorenz's proposal of the innate releasing mechanism had one thing in common with the reflex theory that preceded it. The stimulus produces the behaviour — to all appearances much as the tap below the knee produces the knee jerk. The chain-reflex theory had explained quite well what could be seen up to that time; now there was a new theory which included the old as a special case, given certain simplifications, but also accounting for more complicated events. For a start, the mechanism would allow the released action to appear at the proper time, and the propensity would develop in such a way that the animal would not respond if the stimulus came too early. Next, the threshold level of stimulus (below which there would be no response) might be lowered as time went by with no stimulus arriving. Further, since the propensity was already there, ready and growing, it might eventually be expressed even if there were no proper final stimulus at all, or in response to the wrong stimulus. Without the appropriate trigger, the characteristic behaviour might be redirected into other channels. These were powerful new ideas which Lorenz approached only gradually, although even in Lorenz's first major paper there are signs of an interest in the lowering of thresholds caused by the damming of unused activities.

To be fair to Lorenz's predecessors, many of the elements of the system that he offered were not in themselves new. Even before Mendel the transmission of characters such as birdsong from parent to offspring had been investigated; and descriptions of innate behaviour preceded Darwin. In 1871, Darwin's own book on behaviour, *The Expression of the Emotions in Man and Animals,* listed many activity patterns that appear at birth without the need for learning. A year later, another Englishman was demonstrating genetically determined behaviour patterns in swallows that were reared in a cage too narrow for them to flap their wings, and which were then released at the normal age for flying: the birds flew perfectly well without the usual preliminary exercises or any demonstration of flight. The American H. S. Jennings noted that behaviour could be spontaneous and that behaviour in higher animals often developed continuously from that in animals below them in the same way that their structure did. In 1909, Jacob von Uexküll was talking about the significant world of an animal as being that to which it responded—in other words, the sum of the stimuli which released behavioural responses—the rest going virtually unnoticed. As we have seen, Whitman had begun to think of instincts as comparable to organs, and that their evolution could be studied in the same way. Indeed, it is Lorenz's assessment that one of his own greatest achievements was to put into action what had been laid out by Whitman some twenty-five years before.

The fact remains that this list of prior work can only be put together with hindsight: it was accompanied by much else, from the same scientists and others of high repute, that confused or distracted from the story that was emerging. With the work of Huxley and Heinroth added to this, it still required the appearance of Lorenz for the real start of a more comprehensive attack on the subject; and even then it commanded influence first and most strongly in German-speaking regions only.

Work such as that on *Companions as Factors in the Bird's Environment* which appeared in 1935 as one of his most important observational papers, reached English-speaking readers belatedly and in a severely truncated form. This is not surprising, perhaps, when it originally occupied two hundred and three pages of the *Ornithological Journal of Leipzig*. But some of Lorenz's ideas did appear in translation, such as an article in *Auk* in July 1937, for example, where they were introduced by the comment, 'the doctrine of "releasers" [is] herein set forth as the "keys" to "unlock" or release those "innate perceptory patterns", characteristic of the species and the individual, and which will result in instinctive reactions'. Julian Huxley has written of Lorenz, 'It is to him more than any other man that we owe our knowledge of the strange biological phenomena of "releaser" and "imprinting" mechanisms'.

Lorenz's ideas on releasers were more important than his work on imprinting. Nevertheless, the imprinting technique remains the most spectacular and oft-quoted part of his method. It is easy to demonstrate: an imprinted animal is so clearly attached to the human. Even so, it was to produce its small share of the controversies that later surrounded Lorenz.

Lorenz had first seen the effects of imprinting when he was six, and he subsequently discovered the hard way that in many species the sexual behaviour of the imprinted animal was also altered. 'I remember innumerable hand-reared birds which I quite naively hoped would breed when I put them in an aviary together—and they just didn't but started to court me.' Imprinting was mentioned right at the start of his first major paper, and he found that Heinroth had also had similar experiences. The 1935 'Companion' paper described it as a process of near-irreversible fixation of a drive upon its object, a behaviour pattern which becomes fixated upon that object. This would normally be a fellow member of the species, but abnormally or pathologically it would be some substitute object, and this could be brought about simply by giving the animal misinformation at a crucial moment.

From the objects surrounding it in the hours after hatching, the gosling must find a source of protection and guidance. In the absence of a goose-mother, a human who responds to it will be an adequate alternative or, failing that, it may attach itself to the nearest moving object. Sounds from the substitute creature or object provide reassurance and social contact, and promise the qualities of motherhood. In the experimental situation developed by Lorenz, where he or an assistant has taken on the role, they are provided much as they would be in the bird's natural environment with a true mother. The hatchlings live, sleep, go for walks, nibble at grass and learn to swim in the company of their foster parent. Then after three months the goslings begin to move away from their substitute mother but the family grouping remains stable; the goslings remain loyal to the humans who brought them up, taking encouragement from their presence and responding to their calls made in the birds' own language.

Lorenz found that language—or, at least, each bird's limited range of calls or songs—is one of the keys that may unlock an inborn response. He saw that while greylag goslings simply attached themselves to the first moving object, mallards treated similarly ran away and hid. He fared no better when he offered the mallard ducklings a muscovy duck as the foster mother, but a fat white farmyard duck was readily accepted. Why? Did the voice of the duck, similar to that of mallards, supply the missing factor? Lorenz greeted his next foster-brood with a flood of garrulous mallardese and found himself happily recognised as their mother. Here the mallards must have had an instinctive recognition of the mother's sound but no clear ideas as to her proper shape—except, apparently, that she should not be as tall as Lorenz, a problem which made their upbringing an awkwardly athletic experience. For his neighbours in the village it provided still further evidence of his eccentricity.

For his students, beneath the odd behaviour of their teacher there lay a multitude of lessons to be learned every day. Whether he was caring personally for the many details of his animals' well-being (the success of his work depended on their excellent health), constructing a channel of wooden boards to bring water to the garden duck pond, or a kerosene-heated incubator of his own design, making expeditions that became itinerant seminars down by the river, or giving more formal lectures in Vienna, Lorenz was a hive of activity and an ever-available source of a wisdom that seemed remarkable for a man still in his early thirties. According to Seitz he never vaunted his extended knowledge, but neither was he secretive. Certainly he shared his understanding with anyone who responded to his enthusiasm, and these could include serious amateurs and schoolboys on holiday as well as his students. During a family lunch in the summer dining-room that looked out over the garden, the daily parade of chickens, turkeys, pigeons and peacocks could offer a dazzling array of courtship displays; and Konrad might take the occasion to lecture his fellow diners on such matters as the behavioural differences between the wild and the domesticated animals.

His university lectures were a 'must' in more ways than one. A student who was absent could not fall back on the course lecture notes, for there were none; nor was there any textbook to consult. Lorenz talked of writing his book, but the subject was advancing so fast that the time was never right. In Seitz's view the papers from 1931 to 1941 *are* the textbook, containing the substance of what he discussed in his lectures.

As war approached, in 1938, Seitz's apprenticeship came to an end, although during the early years of the war (before the attack on Russia) he rejoined Lorenz for a further period of research. In 1940 he earned his doctorate with a thesis on the cichlid fish, and later—after the war—he became Director of Nuremberg Zoo, applying Lorenz's ideas on the care of animals as well as he could, and continuing in that role for all but a quarter of a century. But his earliest and continuing claim for fame (with a wide, non-specialist public) is as the photographer of the first films of Lorenz and his geese.

The reason Lorenz wanted the film shot was not merely to illustrate lectures: for these he need only make a few quick sketches on the blackboard augmented by his own vivid mimicry, to characterise the attitudes, movements and calls of a wide range of species. But scientific film was essential if he was to study the sequences of movements that constitute the more rapid, instinctively fixed action patterns, for the film could be studied frame by frame. Stills and film were also necessary in order to establish whether two observers separated by geography or time were describing similar or different behaviour, and to judge any differences precisely. But the reasons for which Seitz's films were originally made is no longer the main cause why they are shown again and again: they are fascinating historical documents enlivened also by both charm and humour.

In the preface to *King Solomon's Ring*, Lorenz tells the delightful story of one lazy summer's day on the riverbank when Seitz set Lorenz thinking that he should write a popular book. Fussing over his camera, Seitz was trying to compose his scenes of the greylags, but the mallards that had also been brought along kept intruding. Lorenz writes, 'I was falling asleep. Then suddenly in the dimness of my senses, I heard Alfred say, in an irritated tone: "Rangangangang, rangangangang – Oh, sorry, I mean—quahg, gegegegeg, quahg, gegegegeg!" I awoke laughing: he had wanted to call away the mallards and, by mistake, had addressed them in greylag language'. Seitz was far too busy to appreciate the joke and Lorenz wanted to tell it to somebody. So, by writing articles and eventually his book, he told it to everybody.

When it finally appeared over a decade later, the book had surprisingly little to say about the geese; he wrote instead of his other animals, and particularly about his earlier love, the jackdaws. In *King Solomon's Ring* he twice promises us a similar goose book, to include, for example, 'the tragic love affair of the greylag goose Maidy . . .', but a further quarter of a century was to pass before, at popular request (mine amongst others), he finally set to work on the history of his geese.

Besides the sad stories of Ada and Maidy we shall have Lorenz's story of Martina, 'the superstitious goose'. As indicated by her having a name and not a number, Martina was imprinted on him. Even as an adolescent goose, when she no longer followed him everywhere, she still went up the grand staircase at Altenberg to spend the night in his bedroom, and flew out of the window each morning. So, as the light fell each day, Martina would wait patiently on the doorstep and when the door was opened to the suddenly darker interior, it was natural for her to go not directly towards the stairs but to head past them towards the light of the tall window at the far end of the great hall. Realising that she had passed the foot of the staircase, she would stop, go back and then up. Repeated day by day, this consolidated itself into a path habit. But as more time went by, the path got shorter and shorter as her track towards the window took her progressively closer to the foot of the stairs where she made a sharp right turn and went up.

One night Lorenz forgot about poor Martina; it became quite dark and she was still outside. When at last he remembered and opened the door, she shot past him and ran straight up the stairs. But on the fourth step she stopped, gave a warning call, turned round and descended to take three paces towards the window. Only then did she feel free to turn and go upstairs again. Back on the fourth step, she shook out her feathers in relaxation, and resumed her scuttling progress to the top . . . the whole performance 'a perfect example of magical thinking'.

After a year of sleeping upstairs Martina became betrothed, and the gander (who had only a number until now) was duly named Martin, after her. Martin had problems, for although a wild gander expects to be bolder than his wife he would never choose willingly to enter a house. Martina could hardly be expected to know this, so it was a fearful Martin who bravely followed his bride over the threshold, to the foot of the stairs and up to the dark terrors of the bedroom, which he challenged with fierce hisses. Every muscle and feather of his body registered the tension of a creature torn between male pride and total desperation. The door suddenly closed sharply behind him and with that, Martin was lost: he flew straight up to the chandelier, dislodging a few pendant glasses and a primary from his wing.

In *King Solomon's Ring*, the last part of this story is already told, with the nostalgic later vision of a wedge of geese flying over Altenberg, birds that Lorenz knew must be his, not only because no other greylags would pass this way (even at migration time), but also because he could see that the spread wings of the second bird on the left lacked that same primary feather.

The definitive goose film has been another of Lorenz's ambitions, and from Seitz's effort onwards there have been several attempts to make it. These have ranged from short scenes to a forty-five minute television documentary in 1973; but Lorenz has never been entirely satisfied by the results. Geese are too personal a love, and the work of others can never express his own special feelings. There is a saying, not entirely true in my experience, that when a camera is turned to nature nothing happens—for wild animals cannot be directed. So the camera is likely to reveal little beyond the commonly observed and predictable behaviour. Certainly it was never there for the magic moments of his memory.

Konrad Lorenz will always be identified with his geese. There is a story—how true or embellished for the telling I do not know—of a Viennese psychologist who visited the Lorenzes one day. Gretl took the man aside and asked, 'Tell me, Professor, can you explain this passion that Konrad has for geese?' To which he shrugged his shoulders and replied, 'It is just a perversion, like any other. . . .'

Chapter 5

Acceptance and Establishment

During the years 1937 to 1940, Konrad Lorenz finally won full recognition for his talents: acceptance by the German scientific establishment. At the peak of his powers, success and its rewards were promised to him. In his father's term he had gained his second glove.

But there could hardly have been a worse time and place, for what was a joke only a few years before was now a political reality. Of those around him, his father was the most ill-equipped to see and point to the growing evils of Nazism, for the old man's conservatism and resentment against the Versailles Treaty predisposed him to any cause that might right the old wrongs. Nazism was on the move, though it would still be a while before these matters affected Konrad directly. And it was in these final years before the conflagration that he produced his finest theoretical papers on the behaviour of animals.

The first of these was essentially a ground-clearing exercise. In *The Establishment of the Instinct Concept* (published in 1937), he examined the ideas of his distinguished predecessors in the study of instinct (the theoreticians Spencer, Lloyd Morgan and William McDougall) and pointed out their flaws. The reign of Purposive Psychology in America under Morgan and McDougall taught that set goals shaped the behaviour that would lead to them. Not surprisingly, Lorenz's dismissive paper had no effect in their home country. In fact, a whole recent generation in America grew up in the shadow of a logical extension of this, the idea that frustration from reaching goals bred violence, so children should neither be set goals they might fail to reach, nor be frustrated from obtaining anything they desired.

In the 1937 paper, Lorenz hardly bothered to mention the errors of Watson and his Behaviourists, dismissing them summarily as obviously knowing little about animals; but in this he underrated their growing influence. Behaviourism with its emphasis on learning from the environment was to achieve a vast following in America in the years to come. (Behaviourism and its relationship to Lorenz's own ideas on instinct will be discussed in more detail in chapter 10.) In 1937 Lorenz still had no complete and satisfying theory to offer in place of those he had cut down (with the moral support and intellectual encouragement of the American, Wallace Craig) although it was clear in his mind that many animal behaviour patterns were truly innate, and he could even see how they were triggered by some pre-set programme. What happened inside the animal that made it all work? Clearly, he needed to know a great deal more about the fine structure of these instinctive actions, and to find out how these were controlled and co-ordinated from within the animal. The behavioural part of this could be tackled by himself and his pupils and friends at Altenberg. The deeper component demanded a knowledge of animal physiology that was greater than his own.

By a mutual interest in bird flight, Lorenz had a friend called Gustav Kramer. (Kramer was interested in how birds find their way about when travelling over long distances, and one of his discoveries was that pigeons use the position of the sun in the sky to provide themselves with a kind of compass.) Kramer was sufficiently impressed by Lorenz's work to set himself to persuading the German academic establishment that Lorenz would provide a valuable source of new ideas, and, indeed, the paper on *The Instinct Concept* was originally delivered at his first lecture to an audience of the Kaiser Wilhelm Society for the Advancement of Science, at the Harnack House in Berlin. In the audience was Erich von Holst, a physiologist whose work Kramer saw as complementing that of Lorenz. It had been Kramer's plan to keep his two friends apart as their science developed along parallel lines, and now that the ideas of each had ripened sufficiently he brought them together, his long-term aim being to persuade the Kaiser Wilhelm Society (which had Research Institutes for different sciences all over Germany) to provide a new institute where the three of them could work together on the mysteries of animal behaviour. It was to be many years before that dream came to full fruition, but the first stage was an immediate success, for both von Holst and Lorenz felt the excitement of working towards an unfamiliar biological concept from different angles.

It was von Holst who had convincingly demonstrated that the physiology of behaviour contained more than the reflexes that had dominated experimental physiology for many years. Pavlov had shown that if a dog's lunch was announced by the ringing of a bell, the animal could in due course be persuaded to produce saliva by the sound of the bell alone: the response was said to be conditioned. This 'conditioned reflex' proved a powerful tool for the investigation of the senses of animals, for the production of saliva was something that could be measured objectively. If a dog was given food after being shown a circle but no food if the signal was an ellipse, physiologists could experiment with the dog's ability to distinguish between the two, to see at what point in the gradual rounding up of an ellipse the dog could see it as a circle and so produce saliva. So widely used and so spectacularly successful was the method that by some it came to be regarded not merely as a tool but also as an explanation of behaviour. As a result, it was supposed that the main role of the central nervous system was to act as the exchange of a complex telephone system. This led full circle to a situation in which nearly all experiments were done with this preconception; some change of environment was made to impinge on the central nervous system and the answer was recorded. 'And the central nervous system, poor thing', commiserated Lorenz, 'didn't get any opportunity in this kind of set-up to show that it could do something else than answering responses.'

By the old explanation, the creeping movement of an earthworm would depend on reflexes. In von Holst's classical experiment, he cut all the connective nerves along which the reflex signal could be transmitted, but still the supposed reflex occurred. Finally, he isolated a single ganglion from the central nervous system together with a segment of muscular tissue in a solution which kept it alive. Then he could show that without any input the central nervous system still sent out signals, 'ping, ping, ping'; it produced a rhythm of its own. Soon it was clear that rhythmic activity can be a basic, internally generated property of pools of nervous elements. The idea of the central nervous system producing such spontaneous signals of its own is so well established today that it seems almost more remarkable that the work to demonstrate this was done so recently—but Lorenz recalls a strong emotional resistance among older physiologists to the idea that the externally triggered reflex was not all.

Lorenz's own work had run parallel to this. He says that he had rediscovered 'what Wallace Craig already knew', that when an animal is deprived of the opportunity of discharging an instinctive motor pattern, not only does the threshold for the release of that action go down, but the creature can become disturbed and begin to search actively for the stimulus which will release the instinctive behaviour pattern. Von Holst realised that he and Lorenz were working on

65

different aspects of the same phenomenon: they were both investigating spontaneity. In Lorenz's case it led him to what has been described as a 'hydraulic model' of instinctive behaviour. In his view, 'energy parcels' produced by the central nervous system and allotted to particular and instinctive motor patterns would continue to accumulate, like the pressure of steam in a boiler, so that sooner or later it had to be let out or the boiler would burst. In its normal discharge the accumulated energy is used up, and to the animal this feels good; indeed, even in its abnormal discharge the animal must feel relief. Among many other observations, that of the hand-reared starling which had killed a non-existent insect in his parents' Vienna flat now made sense.

In his work at Altenberg, Lorenz became particularly interested in anything that could throw more light on the ways in which fixed action patterns work. One approach to the problem was by his pupil, Alfred Seitz, whose job it was to investigate the mechanisms of pair-formation first in his cichlid fish (his Ph.D. paper on this was published in 1940) and later in African jewel fish and other species, in all of which the attraction bringing and bonding male and female together had to overcome the innate aggressiveness that the fish had evolved for the purpose of territorial defence. Seitz had, in effect, to tease apart all of the components of behaviour that were involved in this in order to discover precisely what triggered the 'releasing mechanisms' which made the whole complex system work.

Testing the male fish of his aggressive species with dummies, he found that a realistic model fish produced a smaller response than when he offered a more rudimentary model but with some particular feature strongly emphasised. This might be a special pattern or colour, to mimic those of adult cichlids or sticklebacks, or of motion, as with the jewel fish's way of presenting itself sideways in order to demonstrate a side-beat of its tail. These are all unmistakeable signals to males of the same species that they must fight or flee.

During these studies Seitz had met with one complication which might have upset his results: a very wide variation in each fish's response to the standard threat signals. Eventually this variation was correlated with how much fighting the fish had previously had to do. A male which until then had been living a very peaceful life would 'have a go' at almost anything suggesting the appropriate stimulus, while one which had recently been in many fights required a big stimulus to arouse its interest. To Lorenz this was a classic case of an instinct building up like a hydraulic pressure inside the creature to be discharged by the activity for which it was designed.

But it was the visit of a new-found friend to Altenberg that was to produce the most celebrated study of the detailed structure of a fixed action pattern. Nikolaas Tinbergen was a lecturer who until then had been working largely on his own, studying a colony of herring gulls a bicycle ride away from his department at the University of Leiden in Holland. Because of the similarity of language, Holland had always been close to German scientific traditions and developments; it was therefore natural that Lorenz should be invited to a symposium held in Leiden in 1936. Then, says Lorenz, 'we clicked immediately'; and he returned the invitation.

Like Lorenz, Tinbergen had been an enthusiastic naturalist as a child, with an aquarium that always contained such fish as the Three-spined Sticklebacks so common in Holland. But for the most part he much preferred observing animals in the wild to keeping them at his home. His father, schoolmaster Dirk Tinbergen, doubted that bird-watching would make a suitably serious career for his son, and brought him up with the proper values of a society that placed a premium on 'duty' — leaving the boy with a long-lasting sense of guilt when, after sampling a range of possibilities from photography and farming to physical education (he was a hockey player of international standard), he had finally chosen a career that, whatever its value to society, was so thoroughly enjoyable to him. It was the few delightful months that his father had allowed him to spend at the Vogelwarte Rossitten, the pioneer bird-ringing centre in East Prussia, that had settled the young Tinbergen seriously to the study of biology at university. Then, and again like Konrad, he had married young, and soon swept his bride, Lies, away on an expedition to East Greenland where they lived for a year among the Eskimos and Niko studied wildlife such as snow buntings and phalaropes. But, unlike Konrad, he had tended to keep his scientific papers short — even his Ph.D. thesis ran to a most unusually concise thirty-two pages.

From March to June in 1937, the two men shared a Danube Spring, working together on such matters as the way in which greylag geese move their eggs if they are in the wrong position or become displaced from the nest. If the experimenters placed an egg near the sitting bird she would immediately roll it back into the nest hollow by bending her neck and pulling it towards her with the underside of her beak, and then using the side of her head to steady it. To investigate this

F

further, the two men then made a small change that on his own Lorenz might never have thought of, although in retrospect it seemed simple enough. When the goose had started to rescue the egg they simply took it from her; and now they saw a remarkable thing. With only empty space before her, the bird continued to roll the missing egg into her nest, although without the sideways steadying head movement. The conclusion was simple: there are two separate but overlapping parts of the action, each dependent upon its own cue from the environment. Of these, the head movements in the vertical plane exist as an innate unit of behaviour which, once started, must go on until the pattern is complete, while the head movements in the plane at right angles to that are a series of responses to the feel of the egg going off course.

All this was interesting, indeed intriguing, to watch, but hardly, a casual observer might think, the stuff of a major scientific paper. We may imagine Adolf Lorenz, the near-Nobel-prizewinner, glancing tolerantly at the 'work' of the two young men, but still wishing that his second son were engaged in some real profession: there is no Nobel Prize for ornithology.

Uncharacteristically for Lorenz, their joint paper on egg rolling ran to less than forty pages, but it did contain the writers' clear recognition of the difference between movements such as the orientation of the animal, which are truly reflex-like, and others which involve the release of an involuntary action which is co-ordinated from within its central nervous system.

In addition, they studied the birds' reaction to flying predators by an experimental method that had attracted their attention. A small dummy was pulled along a wire between trees above the heads of the geese. When it moved slowly—backward or forward—the geese immediately became alarmed, but if it went fast they paid no attention whatsoever. To obtain the strongest response it had to move forward at the same apparent speed as a White-tailed Eagle gliding far above. The greylags stared suspiciously at feathers that floated nearby but totally disregarded small birds that darted past; a pigeon gliding in a strong headwind sent the geese scuttling for cover—unless it broke the ominous pattern by flapping its wings. But the eagle-on-wires experiment came to a premature end when the geese began to associate the sight of scientists leaping into trees with the subsequent appearance of the predator, thereby turning an investigation of instinct into a demonstration of conditioning!

Talking of the relationship which developed during that highly productive season, Lorenz comments, 'I am a good hunch producer and Tinbergen is an excellent experimenter who never believed my hunches—and then his verification was most valuable'. And: 'I am an observer, and a lousy experimenter; and Tinbergen's not so good an observer but is a genius at devising what I would call unobtrusive experiments, experiments done on the whole animal. And I would be nowhere without Tinbergen.' In this view the two are the joint fathers of modern ethology.

Tinbergen, too, remembers it as a symbiotic relationship with a kindred soul. He recalls the tremendous vision of his working partner, the flashes of insight from a superb observer and interpreter; and he sees clearly another aspect of their complementary natures: 'He is the farmer, I am the hunter.' Tinbergen, preferring to observe his subjects in a completely natural habitat, could nevertheless experiment by making changes in the animals' environment in ways that would appear natural to them. It has been said less flatteringly that the different methods of the two men reflect the opposites in their personalities, the one domineering and paternalistic, the other self-effacing. But that, too, is an over-simplification: Tinbergen's self-effacement is only one—albeit an important one—of a complex array of personality characteristics between which he can switch with disconcerting speed.

The insights have certainly not all been on Lorenz's side. There are times when you hear an idea that is so good and by hindsight so obvious that you hit yourself on the head and say, 'Why didn't I think of that?' Such was Lorenz's reaction when Tinbergen described his theory of displacement activities—as for example when a goose, frustrated by the restriction to its movement imposed by the slowness of its hatchling infants, paddles her feet up and down on the spot; or when the man watching this curious action scratches his head in perplexity. Seeing Tinbergen's account and explanation of this process of 'sparking-over', Lorenz cried, 'Why didn't I think of that?' Indeed, looking at it in terms of the hydraulic metaphor, Tinbergen had proposed that when the normal outlet was blocked by some opposing pressure, there would be an overflow into some inappropriate channel of activity—which fitted well with Lorenz's own idea of how instinctive drives might work. (In fact, the physiological explanation they envisaged proved wrong in detail, but the underlying concept remains valid.)

Their relationship was at once perfect and impossible. The two together are an explosive mixture in which there are hidden tensions and misgivings as well as intense mutual admiration. Since that first summer they have done little experimental work together and it may be that they could never have worked with each other for more than short periods. Even so, there are few people who can speak to Lorenz as Tinbergen can and few, perhaps none, that Tinbergen respects as he does Lorenz. It was a relationship that survived, in its peculiar way, their total separation by war. The friends were already partly separated within the year, for such were Tinbergen's feelings about the Germany of Adolf Hitler that he could not have visited Austria after the Anschluss.

In Konrad's home there was more mundane talk. Adolf Lorenz's unsuccessful attempt to restore the family's fortunes with the American royalties from his book meant that the family began for the first time to look to Konrad himself as a possible provider. Indeed, for a while in 1938, there seemed a strong likelihood that the prosperous Kaiser Wilhelm Society would create an institute for him. There was also one marvellous moment for Konrad when he first dared to write on the theory of knowledge. He sent a copy of his paper to Max Planck, the great theoretical physicist who had won a Nobel Prize in 1918 and who had first become President of the Kaiser Wilhelm Society in 1930. Planck wrote back saying that it had given him the greatest satisfaction that, based on entirely different data, the two of them could arrive at such absolutely identical views on the relationship between the phenomenal and the real world. Konrad walked on clouds, so delighted was he by praise from a Grand Old Man of German Science.

Final glimpses of this period come from the popular books: Gretl defending flower beds from the geese with wild war cries and a flapping umbrella, and the aging but still vigorous Adolf entertaining a gaggle of geese to tea in the unfamiliar surroundings of his study. Such an honour produced in the birds a nervous excitement that increased the staining power frequency of their droppings. Long after the war years those marks remained as reminders of that day. (The sketch of this event was originally intended for *King Solomon's Ring*.)

In 1938, the great affairs of national destiny marched on with military decisiveness and urgency. And here we must note that although Konrad Lorenz was not so approving as his father of what was going on, he was certainly not averse; nor had he any objection to firm leadership, strength of will or unity of purpose – commendable qualities in other circumstances and at other times.

It is worth considering briefly the emotional atmosphere of Vienna and Austria between the wars that led to the Anschluss, the unification of Austria with Hitler's Germany. It may be suggested that when Austria chose Nazi leadership, its evil nature was clearer than when Hitler came to power in Germany five years before; and yet a majority of Austrians accepted their new leader readily, even with enthusiasm. It was in Austria itself that Hitler developed his anti-Semitism and his understanding of the ways in which it could be used. There is a story, perhaps apocryphal, of the filming of *The Sound of Music* in and around Salzburg, when the locally recruited extras asked how they should act in scenes showing the Salzburgers welcoming the harbingers of the new order. 'Just do what you did before', they were told.

Before the war no Austrian could fail to be aware that, whatever the cause or however justified, his country, once at the centre of the vast 'Holy Roman Empire of the German Nation', had first suffered erosion by the split with Prussia (an effect emphasised by the growth of Germany in the late nineteenth century under the new Prussian empire) and finally saw itself dismembered in 1918. Only the memory of past greatness lingered on. It seemed to many older Austrians that this last abrupt change had been imposed by the arbitrary misfortunes of a conflict to which they had been opposed from the start—or so some could claim, Adolf Lorenz among them. Austria had sought internal reforms but—unlike Germany and Japan in the years following the Second World War—had failed to find new outlets; there could be no comparable redirection of their dammed energies and national pride. The atmosphere was such that even teachers and intellectuals could respectably favour taking part in the great resurgence of the Germanic spirit.

Writing in 1936 for an American readership, Adolf Lorenz praised the intention of President Wilson's Fourteen Points but vividly expressed his disgust for the events that followed. He wrote of 'those men who kindled the World War and then concocted a peace every word of which means hatred'. As they contemplated the way some Americans reacted to them between the two wars, Albert (as quoted by Adolf) said 'Haven't you seen, Father, how unreliable the esteem of your colleagues is? If you are a Teuton, the esteem is lost.'

Today, ethologists such as Tinbergen and Lorenz can point together in agreement to the use that Hitler made of two emotional appeals to man's instinctive responses. One employed the myth of the 'Herrenvolk', where members of one's own race are seen as superior beings; the second called upon the deep-seated need for unity against attack, in this case all the more insidious because the chosen enemy, 'the International Jewish Conspiracy', could be seen to have infiltrated the Germanic peoples at all levels. But despite these pressures within the society to which the Lorenz family belonged, Jews had moved fairly freely within it and there was no barrier to friendship: the Lorenzes had many Jewish friends, including Bernhard Hellmann, Konrad's closest boyhood companion.

In 1937, Konrad had pleased his father by returning to the University of Vienna once again, to teach comparative anatomy and animal psychology as a 'Privatdozent' (a curious institution in Germany and Austria which gives a man the prestige that accompanies association with a university by allowing him to teach, but omits to pay him for doing so). The Anschluss of the following year meant that Konrad's hoped-for institute could just as easily be in the part of the new Greater Germany that had been Austria. This was an idea that must have appealed to Lorenz, who loved his home region, but the war that followed threw all of these plans into disarray. Instead, there followed a long and complicated train of events that carried him far afield, to a Chair of Philosophy at a university in East Prussia.

Lorenz was already aware that his ways of understanding the world around him differed from the scientific method employed by many other scientists. He had become interested in epistemology, the theory of the nature of knowledge and its dependence on man's capacity for perception and his ideological limitations. This led him in turn to the ideas of the eighteenth-century philosopher, Immanuel Kant, who had suggested the existence of 'categorical imperatives', i.e. ethical beliefs absolutely necessary to man. To Lorenz, a categorical imperative appeared to be just another way of saying 'innate'. Kant's ideas lived on in the Baltic seaport of Königsberg, and Lorenz went there to enquire whether the philosopher had anticipated his own discovery of innate releasing mechanisms.

The answer he got was strictly 'no'. But Gretl decided it was time for her husband to read Kant more fully, for plainly there were links to be found with the way he was now thinking; so she bought him a valuable edition of Kant's works. 'With beginner's luck', as he puts it, he quickly found epistemological ideas that excited him, and he wrote to Holst to discuss them.

Von Holst played viola in a string quartet in which Eduard Baumgarten, the Kantian Professor of Philosophy at Königsberg, played the violin — and it was this chance association that provided the key to Lorenz's future. Baumgarten had found his Kantian label both uncomfortable and inappropriate, since he was a pragmatist by persuasion, and at one of their musical meetings he confided his problem to Holst. At Königsberg, they needed a new man who, ideally, should be interested in Kantian epistemology, and should have a practical knowledge of biology. The ideas of such a scholar would fit in well with those of Otto Koehler who was head of the Department of Zoology and a leading light of the University. (Koehler had an unusual combination of interests: he was a brilliant animal trainer, and had made a study of the language acquisition of his own grandchild.)

So through von Holst, with the support of Baumgarten and Koehler, Lorenz was offered the prestigious position of Professor of Psychology at the Albertus University of Königsberg — about as far from home as he could possibly get and still remain on German territory. With the war in Europe at its height, Lorenz travelled to East Prussia to take up his new post on 2 September 1940, to share with Eduard Baumgarten ('each with one buttock') the former chair of Immanuel Kant. Gretl and their children joined him, and of his animals, he took his geese, the fish and a dog. The fish he continued to study, particularly his cichlids whose own lengthy, private battles set him thinking about the roots of aggression. But there was no room for his geese at the University so they had to go to the Königsberg Zoo. They did not settle well, became shy towards their former friend, and were later scattered and lost. Lorenz likes to think that they flew to safety, but this hardly seems likely.

73

The goose summers of Konrad Lorenz's best years had now come irrevocably to an end. In the period up to 1940 he had matured immensely in the power of his work, in his reputation as a scientist and in his prospects. With Tinbergen and others he had revised and extended the theory of animal behaviour by a giant step. They had shown that, as a product of its evolution, each species' behaviour is constrained within a framework that can change only gradually. This evolving structure is rather like a vast and complicated mobile home that is occupied by a succession of individuals who from time to time extend or adapt its uses in response to particular needs of the time. This can never be achieved by pulling the whole structure apart and starting again, as a human designer might, because at no time can the tenant (the developing species) move out. It is therefore more often achieved either by using existing rooms for new purposes or by haphazardly adding new rooms at any point where there is space for them. So long as it suffices as a home—that is, so long as it is good enough for the species to survive in it—the manner in which the house of behaviours was originally put together or has been changed since does not matter. But for any individual tenant there are endless opportunities for losing the way in such a maze, and to finish up by selecting a compartment in a room that holds quite the wrong behaviour. An ethologist would call this 'unadaptive'; for humans we use the word 'neurotic'.

In his study of animals and in the recognition accorded to him, Konrad Lorenz, like his father before him, had found his second glove. Again like his father, he would almost lose it, but in a more complex way. There was not only the trauma of a lost war to come, but also scientific and personal attacks that would close in on him from many sides before he would share with Tinbergen their scientific accolade some thirty-five years later. It is only fair to add that through the vigour of his assertive manner, Lorenz drew much of this fire on himself.

Lorenz's flirtation with Nazism at this time has drawn a great deal of criticism, especially in the last few years before he was awarded the Nobel Prize. There have been suggestions that Nazi preferment won him the Chair at Königsberg, but this is in conflict with the complicated history involving von Holst, Baumgarten and Koehler, all reputable academics. In any case, Nazi aid was hardly necessary for a man of already recognised talent, and for whom something far better and more appropriate would have been found had the war not intervened. By his own account of the matter (as we shall see in the next chapter) the main act that is supposed by his critics to have brought him to Nazi favour actually ended it, and that was before he took up his post at Königsberg. Nazi influence might conceivably have helped him towards his earlier goal of an institute for his animal work, although that is doubtful, since the Kaiser Wilhelm Society was financed by German industry, and the Nazis, for all their power, did not have a decisive say in its running. But active Nazi disfavour could have had the opposite effect and one possibility that has been canvassed is that it was Kramer himself who was the stumbling block in 1938.

Lorenz takes a simple pride in the fact that his first appointment as a full professor linked him with the name of Kant, and his post-war pupils heard many anecdotes of that period. His book on cognition, *The Back of the Mirror* (published after many years of further thought in 1973 and in English in 1976) is dedicated to his memories of Königsberg and to his Königsberger friends Baumgarten and Koehler. When he reached the age of sixty (and before the more recent attacks), his work in Königsberg was described in detail in the laudatio of the *Lorenz Festschrift* and reprinted in the German *Journal of Animal Psychology* (Vol. 20, pp. 385–401, 1963).

The whole of Lorenz's early career makes a logical and coherent progression, with the professorship as its temporary culmination. More than this, it was a further stage in his development, immeasurably strengthening and channelling his powers of philosophical reasoning in a way that is clearly apparent in some of his later papers. And this period of his life would undoubtedly have emphasised his feeling for German philosophical idealism to which he had already been culturally exposed, a form of idealism that tends towards absolutes.

The climate of the period which includes Lorenz's time at Königsberg and the years before offered to all Germans and Austrians the seductive pseudo-philosophy of Nazism. Many weaker but aspiring intellects (and some strong and ruthless ones) accepted it uncritically while it was surrounded by an aura of success. The strongest reaction against Nazism came only from other 'isms', and their proponents tended to disappear from public life. In this intellectually dangerous territory, Lorenz seems to have swum with the stream, for his own purposes. Such is his own contention, and it is in every way consistent with other aspects of the man that he should have done so. It is also characteristic of him that he could have accepted only those parts of an idea which were not inconsistent with his own, while retaining his own strongly held beliefs in their new context.

The most serious issue is that Lorenz allowed the play of these influences to guide his mind and hand to the extent that he wrote a 'Nazified' scientific paper which was published at about the time he got a better job in Vienna, six months before he went to Königsberg. When he realised his error he would have preferred to forget about it (an all too common human characteristic), but unfortunately for Lorenz it did not stay forgotten. When faced with the charge he gave his answer, a reply consistent with his own strengths and weaknesses.

But before we turn to that, there is one further matter to record. It concerns Bernhard Hellmann, Lorenz's closest childhood companion, who was of pure Jewish origin. With his mother, he emigrated from Austria in good time to escape the Nazis, but the country they chose was Holland. After Holland was overrun in 1940, Hellmann and his mother were taken by the Nazis.

Chapter 6

A Private War

Lorenz observed that a dove, when hemmed in, may belie its traditional image of peace and kill its fellow dove without a qualm — no desirable model for human society. The wild wolf, on the other hand, seems an admirable beast: in conflict with a fellow he will spare the opponent who submits, and in general is a model citizen of his own world. So it was not without a certain sly delight that Lorenz put this idea to his readers, in a paper called *Morals and Weapons of Animals*, in November 1935; its substance reappeared later as the final chapter in *King Solomon's Ring*. 'The day will come when two warring factions will be faced with the possibility of each wiping the other out completely. The day may come when the whole of mankind is divided into two such opposing camps. Shall we then behave like doves or like wolves?'

In the thirteen years that followed, Lorenz had plenty of time to ponder those words, although it was not until the publication of *On Aggression* in 1963 that this direction of his interests, his emphasis on the animal inheritance of man, became obvious to a wide audience and gave fresh ammunition to the psychologists and others who were already attacking him. In this scientific confrontation the strategic issue has always been the relative importance of instinct and learning, first in animals and then in man. One major battle has been over the nature of aggression as an instinctive drive, and again, more particularly, as an instinct of man. A relatively minor skirmish was to be fought around his work on imprinting (see chapter 9).

In 1940 all of these conflicts were still to come, though the seed had already been sown. The scene for another skirmish, relatively small in scale but a great deal more bloody, has surrounded the question of 'genetic decay' in domesticated animals and civilised man. The assault here has been more direct and personal, and much of the ammunition has been supplied by Lorenz himself, not so much from what he said — and is still saying — but how he first chose to write on the subject in the opening phases of the Second World War.

Genetic inequality in man is such an emotive issue that it will help discussion of some of the other topics to separate this one and consider it together with the accusations and recriminations that have surrounded Lorenz through the rest of his life, before returning to the other more general controversies. And this story has itself become so complicated that it is worth individually analysing its several strands. These include what he was trying to say, and the way that he said it. Matters have been made worse by a mistranslation: Lorenz was charged at least in part on the basis of an account of his words that is incorrect in one detail that he considers important. So we have also to ask whether the charges still stand when this is put right.

The paper was called 'Durch Domestikation verursachte Störungen arteigenen Verhaltens' ('Disorders caused by the domestication of species-specific behaviour'). It was published in March 1940 in the *Zeitschrift für angewandte Psychologie und Charakterkunde* (Journal of Applied Psychology and Character Study). The existence of this paper has been no secret: it is listed in the bibliography of the second volume of his collected papers. The general thesis is simple. It starts with the idea that the breeding of domestic animals changes not only their physical form but also their behaviour, since this is based on the same genetic system. Domestication—indeed any sort of breeding that is managed by man—has usually led to the emphasis of characteristics of form and behaviour which would make animals less viable in the wild; all of Lorenz's practical work with animals supports this statement. He then argued that the civilisation of man has involved processes that are comparable to the domestication of animals, and he calls the process 'self-domestication'. As a species, we may therefore be undergoing genetic changes that affect our own behaviour and, as so often in the case of animals, this may be for the worse.

Lorenz can offer no proof; indeed, he has no convincing scientific evidence that goes beyond the production of consistent detail and powerful analogies. But if there were any chance that he is right, it would be vital for the entire human race to know as much as possible about the processes involved.

It is a warning that he has expressed in both intellectual and emotional terms. He has a direct, personal and strongly expressed aesthetic response to the sight of domesticated animals that have developed or, rather, degenerated in a particular way. He noted this again in a paper of 1950, 'Part and Parcel in Animal and Human Societies', reprinted in English translation in the second volume of his

collected papers with graphic illustrations. There, he shows us domesticated geese and chickens that are fat, squat, ugly creatures compared
with their clean-limbed ancestors, particularly the wild (greylag) goose
which Lorenz understandably views with particular affection. He
compares the wolf favourably with 'the domestic pug'—and berates
the breeders who permitted the skeletal shortening of thoroughbred
lines so that many of these sad creatures can no longer breathe properly.
The wheezing, waddling lap dog that is the by-product of this process
is a proper object of his derision.

I recall a visit to Seewiesen when Lorenz showed me the subjects of
more recent studies, and I remember in particular his expression of
total distaste for a certain muscovy duck. Some years earlier, his
reaction had been filmed when a human-imprinted muscovy drake
made advances to him. After courting him, with the characteristic
waggling movement where the head and neck are dipped first to one
side and then the other, the drake attempted to copulate with his boot:
the exact analogy of the function of a substitute object that psychologists
call fetishism. Lorenz was not repelled by this pathological redirection
of a behaviour pattern: this has been his accustomed scientific method
and offers a way of studying the unity or integrity of the response. He
would view this aspect of the drake's behaviour with objective dis-
passion. Nor, with his knowledge of the processes involved, would he be
repelled by being the chosen sexual partner of another male animal.
But what he did demonstrate on these occasions was a deep aesthetic
repulsion at the sheer unattractiveness of his suitor. 'Great ugly beast'
was his response.

This reaction expresses sharply a quality that he believes exists in
every man to some degree, an instinctively determined aesthetic
response to the beauty or ugliness of form and behaviour in the fellow
members of one's own species, and the correspondingly similar qualities
(real or imagined) of other species. It is worth paying particular
attention to this idea, as it is important both to Lorenz's argument and
to the misinterpretation of it.

The muscovy duck, he says, is typical of domesticated animals in
that its greed for both food and sex have been enhanced by those who
have created it over the years. Greed in eating is welcome to the man
who wants to fatten animals, and the oversexed animal is easier to
breed in captivity, 'so the domestic animal becomes what is generally
described as bestial'. To the wild originals with their slender grace,
higher intelligence and well-ordered society, all this 'beastliness' is

totally alien. The muscovy is an extreme example, of course, and to show that some domestic animals have been made stupid and obese by deliberate selection proves only that selection works; indeed, there are cases where aesthetically desirable characteristics are deliberately bred in: the Arabian horse is faster and (he says) more intelligent than its wild predecessor, and many domestic dogs are far higher animals than the wolf. But in more cases the reverse is true: as the wild duck is domesticated, it gets fatter, its legs get shorter, the muscle and connective tissues are weakened and the belly sags. In the wild these aesthetically unpleasing characters are selected against quite naturally. Domesticated animals often lose their alertness and sprightliness; but when these qualities are bred for, they may be emphasised in a grotesque manner, as in the racing animal which is nervous tension on legs.

How does all this relate to man? For this is where Lorenz applies the ideas. He sees no reason to suppose that, apart from its greater complexity, man's nervous system is basically different from that of other animals. We do have other instincts, such as moral responsibility: 'Why is it easy to kill a flea, harder to kill a frog, and almost impossible to kill a puppy?' Lorenz asks. In his book, *Civilised Man's Eight Deadly Sins*, he argues that we have an innate sense of justice, the common desire for which expresses itself in a variety of forms that interlock with the various ways in which local cultures have developed; but the common elements revealed by a comparative study suggest that they are based on instinct. We also possess curiosity, a childlike interest in the world about us which is also instinctive, and a potent influence for the progress of mankind.

In contrast to these, he sees today (perhaps even more clearly than in time of war) changes in human behaviour that are similar to the symptoms of genetic decay. These represent a real peril to man's social systems, leading in turn to the danger of collapse of western civilisation itself. All cultures before our own, he notes bleakly, collapsed after reaching a high pitch of civilisation; and he insists that his view is supported by other serious workers who have diagnosed decadence in our own culture. Lorenz attributes this decadence to qualities in the life we live today that are dangerously akin to domestication. Life for the individual is in some ways made too easy, and in others is subjected to social stresses which in animal society could be disastrous. *Civilised Man's Eight Deadly Sins* (published in English in 1974 but written some years earlier) lists those qualities which Lorenz considers most serious;

in general, most will be familiar to those who have listened to the dire warnings of environmentalists, populationists, political ecologists and other social critics over the years.

But one symptom of genetic decay which worries him particularly is infantilism. If the essential childlike curiosity of man is emphasised still further (and it is recently acquired behaviour rather than the older, deeper instincts that would be modified first), could this cause constructive interest to degenerate to the level of eternal childish play? Further: we may suppose that responsibility and altruism are adult properties, but if a man decides that he need not work and that other people will care for him no matter how he treats them in return, this is infantile. 'Let us pray to God that it is not genetic, that it is only cultural. If it were genetic it would represent a very, very great danger.' Such hypotheses are difficult to prove scientifically, but if his extrapolation from animal to man were to prove valid, we should certainly be asking ourselves what action we could or should take.

All this was first discussed in detail by Lorenz in a scientific paper which aimed at being acceptable to wartime Germany's Nazi masters. There was, however, no scientifically respectable need for it to have been quite so clearly angled. (See passages from this on p. 134.)

An early and relatively mild attack on this paper actually quotes a later paper by Lorenz, 'Die angeborenen Formen möglicher Erfahrung' ('The innate forms of possible experience') which was similar in spirit, especially in respect of domestication, but devoid of much of the earlier, overt political jargon. It was published in the *Zeitschrift für Tierpsychologie* in 1943. J. B. S. Haldane, delivering the Huxley Memorial Lecture to the Royal Anthropological Society in 1956, said that from the premise that civilised man is a domestic animal, 'Lorenz proceeded to argue that civilised peoples must inevitably perish "unless self-conscious, scientifically-based race politics prevent it" (p. 302). Such politics are based on "the value of racial purity" (p. 311), "the function of the intolerant value judgment" (p. 308), and other tenets of the *National Sozialistische Arbeiter Partei*'. With no more than this one dig (for in fact the Haldanes and Lorenz were friends), Haldane then stated his own belief that Lorenz was wrong in his basic assumption, and discussed the arguments for and against it on scientific grounds. Although coming down against Lorenz, Haldane obviously thought the question worth raising. But on matters apart from this the 1943 paper contains the most thorough discussion of Lorenz's ideas about the way instinctive behaviour is controlled by external stimuli, and ranks beside the 'Companion' study of 1935 and the 'Instinct' theory of 1937.

But the 1940 paper was another matter, and the controversy over both the manner and substance of that earlier document reached a much wider scientific audience as late as 14 April 1972, when an address delivered in Canada some six months before was published in the American journal, *Science*. The writer was Professor Leon Eisenberg of the Harvard Medical School and (then) Chief of Psychiatry at the Boston General Hospital. Ashley Montagu, an anthropologist and humanist who opposes Lorenz's views on instinct in man, has edited a collection of essays, papers and comments on Lorenz which now contains Eisenberg's lecture, 'The *Human* Nature of Human Nature'. In this Eisenberg states that theories assuming that human behaviour is based on instincts violate the findings of developmental psychology. He wrote of Lorenz's 1940 paper:

In domesticated animals, he argued, degenerative mutations result in the loss of species-specific releaser mechanisms responding to innate schemata that govern mating patterns, and that serve in nature to maintain the purity of the stock. Similar phenomena are said to be an inevitable by-product of civilisation unless the state is vigilant.

Eisenberg then quotes sections from the 1940 paper:

The only resistance which mankind of healthy stock can offer . . . against being penetrated by symptoms of degeneracy is based on the existence of certain innate schemata. . . . Our species-specific sensitivity to the beauty and ugliness of members of our species is intimately connected with the symptoms of degeneration, caused by domestication, which threaten our race. . . .

Usually, a man of high value is disgusted with special intensity by slight symptoms of degeneracy in men of the other race. . . . In certain instances, however, we find not only a lack of this selectivity . . . but even a reversal to being attracted by symptoms of degeneracy. . . . Decadent art provides many examples of such a change of signs. . . . The immensely high reproduction rate in the moral imbecile has long been established. . . . This phenomenon leads everywhere . . . to the fact that socially inferior human material is enabled . . . to penetrate and finally to annihilate the healthy nation. The selection for toughness, heroism, social utility . . . must be accomplished by some human institution if mankind, in default of selective factors, is not to be ruined by domestication-induced degeneracy. *The racial idea as the basis of our state has already accomplished much in this respect.* [Eisenberg's italics.] The most effective race-preserving measure is . . . the greatest support of the natural defences. . . . We must—and should—rely on the healthy feelings of our Best and charge them with the selection which will determine the prosperity or the decay of our people. . . .

Eisenberg then added his own comment:

Thus, it would appear, science warrants society's erecting social prohibitions in order to replace the degenerated innate schemata for racial purity. Lorenz's 'scientific' logic justified Nazi legal restrictions against inter-marriage with non-Aryans. The wild extrapolations from domestication to civilisation, from ritualised animal courtship patterns to human behaviour, from species to races, are so gross and unscientific, the conclusions so redolent of concentration camps, that further commentary should be superfluous. Perhaps it is impolite to recall in 1972 what was written in 1940 but I, at least, find 1940 difficult to forget; indeed, I believe it should not be forgotten lest we find ourselves in Orwell's 1984 for the very best of 'scientific' reasons.

For over a year, Lorenz did not see this attack, although several of his friends urged him to read it. When at last he did, he wrote to me that the great devil he was fighting then, and is still fighting, is the progressive self-domestication of humanity, and his next book, on good and evil, would deal with the same problem. In such a fight it was permissible to recruit *any* minor devil (any ideology promising help). The point of the 1940 paper was that the allegedly 'nordic' ideal of the tall, slender and long-skulled man was due to the projection of a generally human—not racial—aesthetic response. It was not anti-Semitic but anti-domestication; anti-racist even, because it represents domestication as a far greater danger than any possible mixture of human races.

But there is no question, no disagreement about the manner of his saying this in 1940. The statement was clothed in Nazi terminology and Lorenz does not deny it; indeed, he has explicitly said that it was. The reason he gives is that he admittedly and naively hoped to make some propaganda against progressive domestication. But, he insists, he did not make any derogatory remark about any other race of man: it is the putting together of individual sentences out of context that gives this impression.

He then points to a mistranslation. In the paper, he was saying that our aesthetic and ethical sensitivities, both based on innate releasing mechanisms, represent humanity's only remaining protection against progressive domestication, and it is our sexual responses in particular that are sensitive to the symptoms of decadence wrought by domestication. In the paper he then wrote: '*Für gewöhnlich wird der Vollwertige auch schon von sehr gering Verfallserscheinungen an einem Menschen des anderen Geschlechtes besonders stark abgestossen*', which means: 'Generally a healthy and normal person feels a strong repulsion against even slight symptoms of degeneracy in *people of the other sex*' and not (as translated in the Eisenberg address) 'Usually a man of high value is disgusted with special intensity by slight symptoms of degeneracy in *men of the other race*' — which in any case, from the context, it obviously did not mean, even besides any mistranslation of the words themselves.

Put the question explicitly. Was Lorenz attacking Jews? The best clue comes from a passage on page 71 of the original:

The fully superior person ('Vollwertige') reacts against contemporaries manifesting inferior traits by keeping away from them. But this reaction from a superior person is felt by the person with inferior traits ('Ausfallstypus') as extremely galling, and he responds to it with boundless hatred. Where there is a sizeable group of persons with inferior traits, the fully valuable person who rejects them becomes highly unpopular. The high biological and moral demands raised by his unspoiled innate schemata are interpreted as arrogance, his rejection of the inferior-trait persons as lack of social sense. If such a person happens to be old and still close to his peasant ancestry, his rapidly domesticated descendants will reproach him further with sclerotic senility. They join against him in perfect harmony — otherwise not their distinguishing trait — and since they do not limit their reactions to passive rejection, they are much too often able to trick him into a fall. I could draw on my head any number of libel suits if I were quoting by name concrete examples of this process prevalent in all strata of society. The unpopularity of people who insist on a selection of the most decent becomes understandable if we consider that they have taken on themselves a biological function which in prehistoric times was played by inimical nature.

Here Lorenz is clearly not writing of Jews — who in any case were not bringing libel suits against Jew-baiters in the Germany of 1940 — nor of any race, unless 'domesticated descendants' of their own 'peasant ancestors' may be described as a separate race.

So then, is Lorenz a racist? No, by any reasonable standards: and there is no evidence of it in his other writings, while in his own grand-children, his genes have been mixed with those from Prussia (Slavic), Italy, and even Polynesia. But yes, says Lorenz: of course he is a racist 'in a manner of speaking'. The domestic goose is a 'race' of the greylag and an Aylesbury duck is a 'race' of the mallard, and we know his feelings on those races.

Lorenz's explanation, together with an attempt to understand the social and political background of Austria in 1939, reduces the propor-tions of the whole matter to an episode which is small by comparison with so great a life's work. And yet on the question of the Nazi terminology I personally retain a great deal of emotional sympathy with Eisenberg's view. Take out the mistranslation and the Nazi terminology remains. Eisenberg does not speak much German, so he was dependent for his translation on others. There was, in fact, more Nazi terminology in the original than Eisenberg had realised when he wrote his address and his general thesis is, he claims, unaffected by the mistranslation.

The real question then is this: irrespective of any intention on the part of the writer, when is a statement scientific and when is it political? On this Eisenberg quotes Noam Chomsky: imagine a psychologist in Germany in 1936 asking himself the 'scientific' question whether Jews have an acquisitive instinct. Supposing you could actually find a method for objectively investigating the matter, such a study in the context of the time and regardless of the intent of the investigator would have a profoundly political meaning. But in Lorenz's case it was not quite as simple as that. The Nazis had adopted a perversion of 'Social Darwinism' which involved the strengthening of the 'Aryan' stock by breeding men according to rules which they themselves determined. Lorenz was in effect actually asking for a scientific re-examination of ideas that had already been fixed in orthodox Nazi 'philosophy'. He evidently hoped by a little clever manipulation to get his own ideas accepted into Nazism. This would have involved more flexibility than the Nazis were capable of—as we now know and he very soon discovered.

In the correspondence on his paper in *Science* Eisenberg in turn found himself under attack, accused of adopting the 'increasingly fashionable anti-scientific position that scientific investigation be "in the service of man" '. Who is to determine the social relevance of research, the writer asked; presumably some government-controlled authority? Eisenberg was also criticised for implying that Lorenz provided a pseudo-scientific justification for anti-Jewish atrocities, as though the Nazis would have desisted for lack of scientific justification. To this Eisenberg acidly replied that no single act of any individual citizen made Nazism possible, but collective failure to oppose, willingness to acquiesce, and acts of support, all contributed to the holocaust.

Important in their own right, such questions drag the discussion away from what is essentially a purely scientific question—the possibility of genetic decay in humans. Lorenz put it at the wrong time and in the wrong context, although in 1940 he could not see this, for Nazism was dominant in most of Western Europe and seemed likely to remain so. Questions about the future of mankind do not choose their own time to become worth asking, but the result here was that Lorenz has had to spend as much effort in answering the manner and timing of that first pronouncement as in reiterating its substance. At the time when the Nobel award was announced, the Viennese Nazi-hunter Simon Wiesenthal demanded that Lorenz retract his wartime views—and then German press and television became involved. *Der Spiegel* began talking about Lorenz's 'brown past'—a delicate euphemism—and *Newsweek* reported an appearance on Dutch television: 'Occasionally blushing and paling under rough questioning . . . only after repeated pressure did Lorenz face up to the past.' He is reported (*New York Times*, 15 December 1973, repeated in *Newsweek*) to have conceded eventually: 'I regret it . . . I now have very different notions concerning the Nazis.' Again, his choice of words was unfortunate; they could be taken to imply that he had only just realised the wrong of Nazism instead of some twenty years earlier, as was the case.

Another American journal, *The Sciences* (published by the New York Academy of Sciences), joined the fray with an article which had a great deal to say against Lorenz. The author, Wallace Cloud, included a supposed quotation from the 1940 paper, in which Lorenz is said to have called for 'the extermination of elements of the population loaded with dregs'. In fact 'extermination' is a poor translation for the original 'Ausmerzung', while the remainder is a free precis of a longer passage. The result is a grotesque distortion. Cloud's article annoyed several people he selectively quoted; to the extent that the journal went to the lengths of offering all the interviewees a greater opportunity to reply than is usual in such cases. Ashley Montagu (generally rated a strong opponent of Lorenz) was one who had intended to dismiss the relevance of Lorenz's accommodation with the Nazis, but he had been quoted in such a way as to emphasise it; he demanded and received an apology for misrepresentation. The journal itself took the view that the discussion was a proper one, since they were still (two months after the original publication) of the opinion that the Nobel Prize had gone to a man who had 'once written a paper with distinctly genocidal overtones'.

Tinbergen's view of this whole matter should be recorded: by his paper Lorenz did expose himself, although he later changed his view; and he was politically and socially naive. When the two scientists met again after the war they discussed, dismissed, and then dropped the matter for ever.

As for the dangers inherent in self-domestication, Lorenz has no intention of dropping his attacks on those, however much the original context of his statement may have been criticised. The hope that he offers lies in education. Everyone has a sense of values, he says, and we should be educated, and educate our own young in turn, to select a mate not by the size of her breasts but by her intelligence and moral values: the cover girl is a vice of humanity. Education can and should improve our aesthetic taste for what is best in each other, and so effect a desirable form of sexual selection.

This solution would do no harm even if Lorenz's fears were exaggerated or even totally unjustified. In one way or another, we are certainly not the end of our line: man is very rapidly evolving at present. But we have sought without notable success for 'the missing link' between the prehuman and the really human creature; perhaps, and at best, Lorenz suggests, the missing link is us.

The genetic balance within a population, whether in humans, other animals or plants, can shift in two main ways. First, new mutations may be selectively favoured by the environment in such a manner as to increase their chances of survival. Present-day pressures would favour adaptation to a complex technological society rather than to the more natural environments of our forebears. There is potential harm in this only if we are likely to return to earlier conditions; but in the meantime, whatever may happen to the over-fed television watcher and his offspring, human society still values mental and physical abilities; it is even possible that we are slowly changing for the better. Secondly, we have removed the selective pressures against some unfavourable genes that may already exist in our population, or can even arise spontaneously. One example is phenylketonuria, a genetically-determined metabolic disorder which, if left untreated, causes brain damage and mental deficiency. A simple test on newborn babies can now detect it, and highly effective counter-measures allow the sufferers to live a free and normal life. But the inevitable result is an increase of PKU gene frequency in human populations, although geneticists have calculated that even after thousands of generations the effect will still be small.

When there is an increased mixing of populations these processes may speed up. I quote from the final page of E. O. Wilson's *Sociobiology — The New Synthesis* (1975):

> Mankind has never stopped evolving but in a sense his populations are drifting. The effects over a period of a few generations could change the identity of the socioeconomic optima. In particular, the rate of gene flow around the world has risen to dramatic levels and is accelerating, and the mean coefficients of relationship within local communities are correspondingly diminishing. The result could be an eventual lessening of altruistic behavior through the maladaption and loss of group-selected genes. . . . Behavioral traits . . . can largely disappear from populations in as few as ten generations, only two or three centuries in the case of human beings. With our present inadequate understanding of the human brain, we do not know how many of the most valued qualities are linked genetically to more obsolete, destructive ones. Cooperativeness toward group-mates might be coupled with aggressivity toward strangers, creativeness with a desire to own and dominate, athletic zeal with a tendency to violent response, and so on. . . . If the planned society — the creation of which seems inevitable in the coming century — were to deliberately steer its members past those stresses and conflicts that once gave the destructive phenotypes their Darwinian edge, the other phenotypes might dwindle with them. In this, the ultimate genetic sense, social control would rob man of his humanity.

Wilson and his colleagues regard themselves as being of the generation beyond Lorenz in scientific understanding of the biological foundations of behaviour. But that does not prevent Wilson in turn being attacked at exactly the same point in his argument: the extrapolation to man of what is seen as a politically suspect biological determinism. And Wilson is tarred with the same brush as that applied to Lorenz: one attack on Wilson even threw in Cloud's 'extermination' quote for good measure. Nevertheless, Wilson's work has gained many admirers.

Despite his reservations on Lorenz (who wrote on aggression rather than of the reciprocal altruism which interests the new sociobiologists) Wilson's conclusion seems remarkably similar to that of his predecessor. Perhaps, then, we should really be concerned about the possibility of genetic change in human populations as Lorenz warned, whether or not we accept his ideas on 'self-domestication'. Wilson bases his statement partly on the work of Haldane published in 1932, work that should also have been known to Lorenz. Let us turn again to the paper in which Haldane considered Lorenz's propositions.

In 1956, J. B. S. Haldane accepted that selection pressures on civilised man have been sharply reduced. Unlike man, domestic animals are deliberately inbred: they become highly specialised in the characteristics we desire, but by the inbreeding itself usually lose many of the important characteristics of their ancestors. In contrast, man as an animal species is characterised by his remarkable lack of specialisations, and can turn his hand to almost anything: 'No other animal can swim a mile, walk twenty miles, and then climb forty feet up a tree. Many civilised men can do this without much difficulty. If so it is rather silly to regard them as physically degenerate.' Then what about the degeneration of behaviour in domestic animals that Lorenz pointed to in his 1940 paper? Haldane picks up one particular aspect of this, the reduced intercommunication within domestic species. In contrast, communication in man is hypertrophic: we speak, write, gesticulate, draw, perform rituals. He adds that an ethologist could describe religion as a vacuum activity of communication in which human beings communicate with non-existent hearers.

However they may stand up to close examination, Lorenz's propositions on man have stimulated further research into the nature and result of domestication in animals. At the University of Kiel, for example, Professor Wolf Herre has studied the way in which conscious or unconscious selection by man over many generations produces changes in the nervous system which have occurred at the same time as changes of behaviour. Herre found that there was a marked reduction in the size of the brain in domestic animals when compared with their wild forebears. In cows, horses and pigs this reduction may amount to thirty per cent. Comparative anatomical examination shows that this particularly affects the regions of the brain concerned with sensory perception and the control of movement: these are easier to find and study than is any direct relationship between behaviour and particular parts of the brain. Reduction of brain size is a result of genetic change, demonstrated by experiments in which the domesticated and wild forms are interbred and the hybrid forms examined in both their anatomy and their behaviour. Lorenz produces no evidence of any comparable trend in man.

However civilised man may compare with lower animals, domestic or wild, there is one great difference between our present civilisation and those that came before, and fell. Lorenz, at his most optimistic, argues that we in our civilisation have sciences, and have thus created at last a tool for man's objective examination of man and hence can study our own nature; we are also the first and only species to have such a capacity. In giving us the ability to reflect upon our own condition, science allows us to see the signs of collapse before they happen. He concludes: 'We might just be in time to stop the apocalypse. But it will be touch and go.' Most scientific opinion today, however, follows Haldane: if the apocalypse comes, it is thought unlikely that it will be due to genetic decay.

In 1974 I asked Lorenz if he thought that the effects of man's self-domestication would ever be studied to his satisfaction. He hoped so, but 'if I have changed my opinion a little it is because cultural decay is so much faster than genetic decay, and so is more imminent. And that's a thing I realised only lately, how much like a living system a culture is. The only things that still tend to keep humanity on the right path are the emotional feelings for values—something that is rapidly getting lost in technology.' (We shall return to this in chapter 12.)

'The 1940 paper tried to tell the Nazis that domestication was much more dangerous than any alleged mixture of races. I still believe that domestication threatens humanity; it is a very great danger. And if I can atone for the (retrospectively) incredible stupidity of having tried to tell the Nazis about this, it is by telling the same obvious truth to quite another sort of society which likes it still less, that same truth!'

Whatever else they might have been, the Nazis were not fools; it can hardly have mattered to them one way or the other what was said by an eccentric Austrian goose fancier. Lorenz says they made no trouble for him but simply ignored him from that time onward. If he had ever had any chance politically he spoilt it completely with that paper. And so he went to Königsberg, in a political vacuum. He could still get his scientific papers published, amongst which were the 1943 paper and his main (1942) paper refuting 'Vitalism' as an explanation of behaviour. 1941 had seen the publication of 'Comparative Studies of the Motor Patterns of Anatidae' (dealing with the family of birds that includes ducks and geese) which catalogues in detail certain easily recognisable acts performed during courtship by each of the various species he had studied with such loving care at Altenberg. This was so long it originally had to be published as a special edition of the learned *Journal für Ornithologie*. In it, he seems at first to be engaged solely in recording a list of behavioural elements that distinguish one bird from another, as many naturalists had done before him. But he then applies the comparative approach that he sees as the proper scientific end of such studies, and constructs evolutionary trees on the evidence of behavioural patterns, using the all-important similarities as well as the differences of behaviour. It remains an excellent example of the nature of much of his science until that time.

The family spent one full academic year at Königsberg. Adolf joined them for the winter, when Gretl gave birth to a second daughter, Dagmar, and he returned to Altenberg in the Spring. Konrad thoroughly enjoyed the sessions of the Kantian Society which lasted long into the night, and at his Institute for Comparative Psychology there was work to be done with fish (it was at this time that he arranged for Seitz to have leave from the Army to continue his studies of bonding—and with that, aggression).

And so, having set the scene for his later personal war, Lorenz could work on for a while in academic seclusion, well away from Germany's main theatres of aggressive attack, which now lay far to the south. But the Russian war was soon to begin.

Chapter 7

The Eastern Front

The war in the East and its aftermath forms a watershed in Lorenz's career which was opened out by a gap of six years, first as an Army doctor, then as a prisoner-of-war. Because they seem to interrupt the flow of his successes, these episodes could easily be glossed over and forgotten in the telling of his life's story. But much of what happened to him is typical of his character and yet more was formative in its continuing development.

After sweeping victoriously through the Balkans to the Mediterranean, on 22 June 1941 Germany advanced over the frontier of divided Poland and into Russia. There were now air raids on Königsberg, and it suddenly became a perilous place to live. With no political preferment, Konrad was called up for war service. Although he had worked in the accident surgery in Vienna to prepare himself, he thought his medical knowledge too poor for work as an Army doctor. So when he reached the section on 'special skills' in his draft papers, he merely entered motor-cycling and teaching, and actually got as far as becoming an instructor in a motor-cycle squadron before the system caught up with him and he was posted to a relief hospital for the military at Poznan, in Poland.

There Konrad worked as a neurologist in a psychiatric unit, an experience that was both terrible and fascinating. He found himself treating a young man in his first attack of schizophrenia and had the shattering experience of becoming his confidant, with the attendant danger that at any time the patient might turn against him. Of his special treatment centre for hysteria, his memory carries scenes 'too horrible to tell. They are ghosts . . . fiends! The most awful thing you can imagine is a human being which has lost its humanity'. (He stresses the 'its' with a grimace.) Yet he considers himself lucky to have had first a year of battling his way through Kantian philosophy, and then these two years in a psychiatric clinic. He would never have chosen so ugly an experience: it was fateful that he should have been so firmly steered towards it. This period added directly to his knowledge of human behaviour. In many of his cases the answer to Bernhard Hellmann's question, 'Is this as the constructor intended?' had to be an emphatic 'No!' He could also look back on them in the light of further questions suggested by a Leeds psychiatrist, Ronald Hargreaves. First, what is the normal survival function of the process here disturbed? Then, what is the disturbance, and, in particular, is there an excess or a deficiency of the function in question?

Movement of civilians without good reason was now very difficult, but when Königsberg University asked Gretl to give up their flat for someone else, she seized the opportunity to justify taking her children and their belongings back to Austria, where they all arrived safely with the exception of the packing case containing Konrad's most valuable papers.

Meanwhile, for joining a protest against the dismissal of Jewish professors in Holland, Tinbergen was taken and held in a Nazi camp as a hostage. He does not like to talk about it, but says simply that by Nazi standards the treatment was not bad, although on two occasions parties of the hostages were shot in reprisal for acts of sabotage in Holland, and escape was considered impossible because they were told that for every one that got away ten Jews would be shot. Koehler and Lorenz came to hear of his arrest and wrote to Tinbergen's wife offering to intervene. Lies could not consult her husband, but she refused on his behalf because he was there as the consequence of a political and ideological clash with the Nazis. Tinbergen later agreed with her decision: although the risk in staying in jail was great, as a self-respecting Dutchman he felt that he really had little choice. Both Koehler and Konrad respected his wife's request of 'please, hands off', but Tinbergen evidently did not know that intervention was, in fact, made, by Eduard Baumgarten at his friends' instigation. He spent a little over two years as a hostage, and when released, he got back across Holland to Lies and joined the Resistance movement.

Baumgarten sought also to secure the release of Hellmann, but in that case was unsuccessful. Neither Bernhard nor his mother were seen again. Lorenz believes they were gassed.

Painful though it must be, Lorenz still speaks or writes of him when the occasion is appropriate. In *King Solomon's Ring* there is a brief reference which will have meant nothing to most of his readers, to 'my tragically deceased friend Bernhard Hellmann who was able to duplicate, at will, any given type of pond or lake, brook or river'.

I asked Lorenz when it was that he first came to realise the evil of Nazism, to which he replied, 'Surprisingly late'. It was in 1943 or 1944, near Poznan, that he saw transports of concentration camp inmates (in fact gypsies, not Jews) and with this evidence of his own eyes he at last 'fully realised the complete inhumanity of the Nazis'.

On the occasion of his ninetieth birthday, Adolf Lorenz broadcast on German radio, although Konrad did not hear it. By then he was stationed near Vitebsk in White Russia. There, as a field surgeon working in a concrete bunker close behind the front line, he saw in horrifying detail what man could do to man in the name of war, where 'the highest of all the vertebrates perpetrated mass mutilation on members of his own species'. Then the Russians swept westward, around the Germans at Vitebsk, and on 24 June he was captured. As he tells it, that story is a small saga in its own right.

In the confusion of the battlefield he had become separated from his party and found himself in the midst of a panic retreat. To avoid being swept away by it he turned and strode out in the opposite direction. 'The most awful thing you can imagine is great masses of men in a panic, with starting eyes, bulging eyes, completely blind, only running, running, running. And I felt very proud of going in the opposite direction. That's what my reaction was, as the only means of not being sucked into the panic.' He looked round, saw a man following, and waited for him; and gradually they collected together about fifty men, mostly sergeants. Cut off by the Russian advance, they tried to break through. With Lorenz to the fore, running and yelling, they took a Russian trench—but there they stopped. The courage of the soldiers ebbed and they refused to go on; so Lorenz left them, to try and get through on his own.

At one point he drew the fire of a Russian who was posted to prevent the escape of German soldiers: 'He did me the honour of shooting at me personally with a light field gun while I was zig-zagging up the slope of a valley opposite.'

During the night, he had to cross a road where the Russians were marching; he removed his cap, tore off his insignia and marched with them: 'I joined the Russian Army.' After a while he went to the side of the road and busied himself by some bushes; when nobody was looking, he leapt into them. To work out which way he should go, he searched for a Russian battery to observe the direction in which they were shooting, and he followed that line until he came to two trenches that were firing at each other. Passing around the first, he ran towards the other shouting, '*Nicht schiessen, deutscher Soldat!*', and the men in the trench ceased fire. He took a deep breath and walked towards them—only to discover that the heads raised up to greet him wore Russian helmets. Russians had been firing at Russians.

A wave of intense disappointment swept over him and he ran again. A bullet hit his left arm, but still he got away. He made a detour round that trench, found a field of wheat and hid in it; then, exhausted, he fell asleep.

While he slept the Russians came, and they woke him saying '*Komm heraus, Kamerad*'. 'They were quite nice about it', Lorenz commented. One of the soldiers who had been in that last trench recognised him and told him what had happened: a Russian encircling movement had drawn in to kill all the Germans in the middle until finally the two sides of the pincer were shooting at each other. It was into this he had walked: he was lucky to come out of it alive.

Until that time he had always felt a slight inferiority in the face of military men and their courage. But from the three days of his own war he learned something new: anyone can have courage. There is no need of any special respect, he says, it is quite easy.

As a captive, he was sent behind the lines to a camp where he was the only medical man among the German prisoners. Despite the shot through his arm and some inflammation, he was able to operate on the wounded. A Russian surgeon was refusing to make amputations, and Lorenz thought he intended to leave the Germans to die of their wounds, until he learned that in Russia such wounds would have healed satisfactorily. The Russians, it seemed, had a much stronger resistance to infection than the Germans.

Lorenz worked on, operating on case after case until, utterly weary, he heard that another two hundred wounded had arrived. He stumbled out to see, and there, sitting quietly before him, were all the medical men with whom he had worked before his capture, many in better condition than he. His exhaustion finally overcame him and he fainted. When he came to, he found that he himself was on the operating table, and as the surgeon worked on his arm Lorenz was told to count aloud. He did not, but talked instead; in fact he gave a lecture. Afterwards he remembered nothing of what he had said but was complimented upon it: it had been very interesting!

Lorenz's public war was over. Now began a long interlude, as he was despatched to prison camp near Erivan, at the foot of Mount Ararat in Soviet Armenia.

In Austria, when her husband was reported missing, Gretl retreated still farther from the action. She took her children to a valley in Vorarlberg, the province that shares a border with Liechtenstein and Switzerland. But Adolf in his old age could not bring himself to leave Altenberg, not knowing whether he would ever see it again. So for the three winters from 1944, Thomas, Agnes and Dagmar lived with their mother in the small town of Schruns, where the war was only a remote echo. Gretl's qualification as a gynaecologist was sufficient to keep them comfortably, and their distance from the fighting eventually saved Thomas from conscription to the bands of bewildered children that were Hitler's last desperate defence against the victoriously advancing Allies. Dagmar passed through the most formative years of her life with her father absent at the war or in Russia, but she never lacked a male to play that role, for it was well filled by Thomas. And so it was that brother and younger sister must always be closer in their bond than were the father and his third child.

In February 1946, aged ninety-two, Adolf died. The family attributed this not just to old age but also to his lifelong complaint of a 'delicate' stomach, and his difficulties with the unsuitable foods of the war and the subsequent occupation. In the last dozen years of his long life he had seen a resurgence of 'the Germanic spirit' grow to dominate most of Europe and then wither beneath unrelenting Allied attack. He had witnessed the last blind conflicts of a war that in its final bloody phases reached the very woods around Altenberg. But there was one consolation in his ruined world, for before he died he heard that his younger son had survived the war in Russia and was safe, a prisoner.

Adolf's property was divided between his sons. Albert's third wife, Marigen, had no liking for Altenberg, and so (with Albert) received the other properties and such money as there was, while Konrad's wife could come back and take over the house her husband loved and keep it for his return. The building had suffered little from the war, although the roof was leaking (due, it was discovered later, to the jackdaws), and rainwater ran down one wall of the big hall: the stains are there still.

But in Altenberg Gretl no longer made the family's living as a doctor, becoming instead a farm manager. Some of the lands put together from her parents' holdings and those of another family came under Gretl's care. Part that was formerly a tree nursery had also been given over to food crops, so the family itself always fed well, and their role as producers gave them a little extra status with the Russian authorities. Gretl earned Konrad's deep admiration for her courage during the occupation. At first the Russians accused her of sabotaging her own water supply—the well in the garden was out of order—but as her husband tells it she would have none of that: 'Without knowing it she treated them just right. She dressed in dirty overalls and she shouted at them, which was exactly the right way to treat the Russians. They were very kind to her', he added. Gretl agreed with that assessment. She became friendly with some of them who turned to her for medical advice. Nearby lived a Russian general whose wife was neurotic, complaining that she could neither eat nor sleep and was afraid to leave her house. 'Gretl had her tame in no time', commented Konrad.

One pair of Gretl's midnight patients were soldiers who had got drunk and inflicted deep knife wounds on each other. Sheepishly, they sought treatment from her for they did not dare to confess their misbehaviour to their own doctor. Another time, some soldiers came and demanded 'in the name of the General' to put a large, heavy barrel in the cellar. Gretl was suspicious, but 'you didn't argue with the Russians'. Then, one night, she heard noises and went downstairs to find that the men had broken in to retrieve their booty. With some assistance from her nephew Georg, Gretl imperiously arrested the intruders and held them prisoner until morning. By then they were contrite and anxious to repair the damage, pleading with her, 'Please, please, do not tell the General!'

In his prison camp the Russians gave Konrad no reasons for hostility either. 'I was never in a really bad camp', he told me. 'If you had the bad luck to be in a camp where the prisoner kapo was a criminal – which he often was, because gangsters have the knack of forming a state within a state and getting on top—and this was combined with a dishonest Russian officer, then you likely died of hunger. However, if the prisoner kapo was an honest man and the Russian officer too, you could live in perfect health.' The farther the camp from Moscow the greater the chance that conditions would be bad. In Armenia Lorenz was over a thousand miles from Moscow, but it seems he was none the worse for that. To his knowledge, the Russians were never cruel to their captives. He heard later reports of 'awful things' from some American and particularly French camps, but in Russia there was no sadism. Lorenz was never persecuted and there was never any hostility on the part of the guards.

One report, albeit at second hand, came from a German ex-prisoner who had been in the same camp. By his seniority and popularity, it was said, Lorenz had authority among his fellow prisoners and was respected by his captors.

One exploit in particular added much to his reputation. Seeing him about to catch a big tarantula spider, a kindly Russian guard warned him of the danger, saying that it was very, very poisonous. Lorenz promptly picked up the spider and, safely gripping the head and thorax, bit off and consumed the fleshy abdomen. This sent the poor guard 'running and screaming into the steppes of Kazakhstan', according to Lorenz. Before the war, Seitz had noted his concern that the food of his insect-eating songbirds should be tasty and succulent: 'One could see a certain joy or satisfaction when the quality, consistency, and also the aroma were correct for the little feathered consumers'. And Thomas, too, recalls his father eating parts of the big june-bug with satisfaction. Lorenz loved to astonish an audience, but he could also demonstrate a simple practical point: there have been reports of expeditions whose members had starved to death when there was food all around them. In particular, the eggs or egg-mass of even the most poisonous creatures is safe and good to eat. The spider eating was not all display, declares Lorenz; at Erivan he really needed the extra nourishment provided by its protein.

But despite their poor diet, the health of his fellow prisoners was reasonably good, so his medical duties were light. There was little opportunity for systematic animal observation, although he was able to raise a few birds, and fleas constantly offered themselves to his involuntary attention. Indeed, the fleas could be studied by touch alone, on the skin beneath his shirt. Some of them raced around in circles, and he was able to verify that such fleas were male, and that a female was to be found at the centre of their attention — as also occurs in the courtship behaviour of some flies.

With little else to occupy him in the camp, Lorenz had plenty of time to think and to write. He turned again to philosophy, and particularly to epistemology, the theory of knowledge. There was no writing paper to be had, so he made do with a substitute prepared from cement bags; nor was there ink, so he bleached the words into the coarse surface with a solution of potassium permanganate. In this fashion, a covert manuscript was gradually built up.

By the time he was moved from Erivan to Krasnogorsk, a precinct of Moscow, the existence of his writings had been discovered, and the question now was whether he would be allowed to carry these documents home with him. He was told that the censor must read them, so it was all typed out. This he did with the Russians' aid, and sent the copy off. Then he waited. His repatriation was due in time for the Christmas of 1947, but the necessary permissions had not yet come. His family also waited through that Christmas, in vain. As more weeks went by, Lorenz became increasingly nervous, though occupying his time with more writing. At last, he was sent for.

'Can you give me your word of honour, Professor, that this manuscript includes nothing that is not contained in the typescript you sent to the censor?' the Commandant asked him. Missing the point entirely, Lorenz began to explain that he had changed one chapter, much shortened another, added a third. The Commandant smiled as he stopped him. 'No, Professor, don't misunderstand me; what I am asking is whether your manuscript contains nothing more than your scientific book?' And on a promise of that the two men shook hands.

The Russian told the convoy officer that Lorenz was not to be searched, and that this instruction must be passed by word of mouth to the next convoy officer, and so on. In February 1948 Konrad Lorenz came home at last to Altenberg, ravenously hungry but otherwise in good health and vigorous spirit, carrying a rucksack full of manuscript, a makeshift cage with two birds, and a great number of bed bugs. The infestation of bugs was vigorously and successfully attacked by Annie Eisenmenger, a friend of the family who was later to assist with many of the marginal drawings for Konrad's popular books. The extreme shortness of sight that allowed her to sketch insects directly by the unaided eye now allowed her to peer closely at bedlinen and mattresses, scanning the seams where the pests accumulate. Rarely could a bug escape such penetrating attention as that of Annie Eisenmenger, illustrator-cum-exterminator.

Much had changed in Konrad's absence. There was a daughter to whom he was a stranger and there were none of his prewar animals to greet him — with a single great exception: looking up to the parapets of the Altenberg house he could find one remaining, faithful group among his free-flying birds — the jackdaws. Yet their continued presence both as friends and as the objects of his scientific curiosity owed much to a half-chance event which involved the celebrated English playwright and novelist, J. B. Priestley.

Following the war, the writer and his wife Jane had been invited to visit the Soviet Union, and on their way back across Europe, they stayed in Vienna where Priestley had royalties due to him that were 'blocked', and so could not be used outside Austria. But after the exhilaration of Russia this gloomy, divided city was a let-down, and as Priestley could neither take his money with him, nor wished to spend it by staying, his wife found a deserving use for it. Jane Priestley's consuming interest was birds (she has since married the ornithologist, David Bannerman), and when, through friends, she heard of the problems at Altenberg she arranged for the royalties to be transferred to the Lorenzes. Gretl then managed them and every two months accounted to the Austrian Academy of Sciences through which the funds had been given.

As Gretl tells it, her desk at that time had three drawers for money, the first from the Priestleys for Konrad's work, the second for the business of the farm, and the third for their own livelihood—and that third drawer was always empty! However, the jackdaw-damaged roof could now be put in repair, and Konrad's gratitude (which he still feels strongly) was expressed when *King Solomon's Ring* was dedicated 'to Mr and Mrs J. B. Priestley, without whose timely help jackdaws would not—in all probability—be flying around Altenberg any more'.

Chapter 8

The Second Spring

Lorenz had returned from his enforced rest urgently ready to start a new life and to create for himself a 'second Spring'. His first problem was the question of a suitable job, for it was now so long after the war that the reconstruction of institutions was well under way. Attempts had been made to keep spaces open for him in the jig-saw, but the process of 'fitting in' such an important and influential man was bound to be awkward at this late stage. So there was nothing, no professorship to dignify his name; and with the translation of Königsberg into Kaliningrad, Russia's westernmost strategic seaport, his earlier job had gone for ever.

Much else was also new; but not, as yet, the emphases in scientific thinking. Physics, until then the dominant science of the twentieth century, had gone to war, and its most fearful creation, the atomic bomb, had brought a shadowed and haunted peace even for the victors. In Germany as in other countries, the physics that had confirmed power in war was still looked to for the foundations of the expected technological solutions to man's future problems. For the time being it was still the physical scientists who called the tune in the higher councils of German research.

After the suicide of the previous President of the Kaiser Wilhelm Society in 1945, eighty-seven-year-old Max Planck hastily took on the job to hold it against an improperly appointed claimant in Berlin. Planck, whose peaceful genius had helped to assemble man's knowledge of the workings of the atom, hoped to hand over as soon as possible to Otto Hahn, the German pioneer of nuclear fission who had split the uranium atom in his laboratory just before Christmas 1938 (giving the Germans a lead which, by accident or design, was not followed through by Hahn and his fellow scientists). For his discovery, Hahn was awarded the Nobel Prize for Chemistry at the war's end, which confirmed his claim to the Kaiser Wilhelm Presidency when Planck lay mortally ill in 1946.

When, soon after, the Allied Control Commission banned the Kaiser Wilhelm Gesellschaft, Hahn pleaded his case with the British, who agreed to allow its substance to continue within their zone, but not its former title. So Hahn turned to Planck for the great man's permission to use his name in this cause; and when it was granted, the way was opened for the phoenix 'Kaiser Wilhelm' to be reincarnated as 'Max Planck', and Otto Hahn could be the first President of the Max Planck Society without further opposition. Planck died in October 1947, and in February the following year, the month of Lorenz's return, the Americans also recognised the research society, and the 'new' Gesellschaft was formally inaugurated at Göttingen.

With these events, the end of one era signalled the birth of the next, during which the pre-eminence of the physical sciences was destined to be challenged by biology, Lorenz's own science. But all Lorenz could see at that time was that his former intending bene-factors, reconfined within only part of the new, reduced Germany, now lacked the power to create any new institute anywhere, let alone in his own beloved Austria.

One group of Austrians did welcome him back with open arms. Their leader was a remarkable idealist about ten years younger than Lorenz, called Otto Koenig. As a boy, Koenig had run away from school and begun to make a name for himself as a wildlife photo-grapher and self-taught scientific investigator of the behaviour of the Neusiedlersee animals and birds. Just like Alfred Seitz, he met Lorenz and formed a deep admiration for him. With a number of younger, biologically-minded friends, Koenig had appropriated the site of the former Viennese anti-aircraft command centre on the Wilhelminen-berg, a hill where the western suburbs of the city reach up to the Wienerwald. They reconstructed the huts and tacked on aviaries, built a small dam for a pond, bought or caught their birds and other small animals, and set themselves to rear and study them, all between visits to the public library to read Lorenz's papers, and writing for their own small periodical, *Die Umwelt* (literally, 'the Environment'). Even today they still wear a dull green battledress uniform and live in conditions that are rustic, to say the least.

At the start, their whole irregular enterprise was financed out of the proceeds of a popular book on the Mediterranean and its shores, made up from letters that Otto had sent home from the war in the south, the photographs he had managed to bring back, and further illustrations by his fiancée Lili. Soon their efforts were rewarded by official recognition and academic reward; Koenig was given the title of professor, and Ph.D. students flowed in—about thirty-five by the time of Lorenz's return. The group of able students gathered around Koenig at the Wilhelminenberg Institute included Irenäus Eibl-Eibesfeldt, Wolfgang Schleidt, Heinz Prechtl and Ilse Gilles (later Prechtl), all impatiently awaiting the return of Lorenz, their as yet unseen teacher. Schleidt, who had started by raising mouse families at the Wilhelminenberg, has remained an animal ethologist but several of the others would later parallel Lorenz's interest in human behaviour with their own direct studies.

Koenig is now known not only for the heron work that he had taken over from Lorenz but also for his studies of the colourful military uniforms that man has until recently worn for the purpose of territorial conflict. Eibl-Eibesfeldt has since been given his own department to study human ethology, where he has attempted to identify innate expressive gestures common to human cultures which also appear spontaneously even in children born both deaf and blind. Prechtl had a head start on the others, for he had studied medicine and become interested in the behaviour of children, and subsequently in the correlation between brain and behaviour. In order to make a physio-logical study of this, he turned to zoology, choosing lizards as his main experimental subjects. One of his early papers records an experiment involving both animals and humans: he showed that infants between two and four years will grasp a snake that is held writhing before them, but if they see the same creature wriggling on the ground they will edge away. He concluded that this human response to a snake moving in its normal way must be innate. Within a few years, Prechtl had leapt back across the gap between animal and human studies to build a reputation on the investigation of the childhood development of the human brain.

Some of Lorenz's first lectures after his return were given in the open air, walking around the Wilhelminenberg. 'All this was originally built for Lorenz', says Koenig today, looking about him at the buildings and achievements of his institute; but that Lorenz himself might have joined him there after the war was a romantic dream. There was no money for the sort of research that was now Lorenz's due: to have joined this 'boy scout' group would have meant reliving his own youthful struggles all over again.

For a while there was talk of a professorial chair at the university in the Austrian city of Graz, but this was subject to a 'Proporz', a curious political institution that had arisen in post-war Austria, a sort of gentleman's agreement between the Catholic and Socialist parties that each had a right to sponsor a proportion of the academic positions available, into which went their own men. Lorenz, by his own political history a member of neither group, was excluded by both and the job at Graz came to nothing: indeed, the chance of any professorship in Austria seemed out of the question for as long as political sponsorship continued.

One of his first post-war visitors from England was the Cambridge zoologist, W. H. Thorpe, who had come to consult him on who should be invited to a Conference of the Society for Experimental Biology which in 1949 would consider physiological mechanisms in animal behaviour. In Cambridge for that meeting, on their first visit to England after the war, the Lorenzes stayed with Thorpe, and it was in his house that Konrad again met Tinbergen, and the two discussed and set aside their political differences to renew their friendship. Niko was unhappy to hear of Konrad's difficulty in getting re-established, and predicted that in the end he would go to America, or come to England like himself. They agreed to meet again in Leiden after Tinbergen had visited Oxford to see Alister (later Sir Alister) Hardy about the job he was to start in September. Thorpe drove the Lorenzes to the new goose-keeping station at Slimbridge in Gloucestershire, and to stay with the Priestleys in their house on the Isle of Wight.

At Slimbridge, where Peter Scott had recently laid out grounds by the River Severn for his Wildfowl Trust, new possibilities opened up for Lorenz's future. Supposing an associate professorship could be arranged for him at Bristol, these wildfowl grounds on the banks of the Severn might make an ideal base for renewed greylag studies. Lorenz's reputation had preceded him, and now Scott could see the human substance at its centre. He recalls with particular pleasure Lorenz's capacity for mimicking animal behaviour. Describing how a baby wood duck climbed in and out of its nest hole in a tree, Konrad would suddenly jump up and demonstrate, instantly and recognisably becoming the wood duckling itself by making little jumps at the foot of the wall, and running up it with his hands.

Later, when the offer of a place at Slimbridge was all but finally accepted, Lorenz received 'a very interesting offer' from the Max Planck Society which, in order not to lose him completely, offered to pay a salary with no strings attached — simply for him to stay and

work at Altenberg. He wrote this to Peter Scott, and Scott wished him
well in his change of plan.

Back at Altenberg, Lorenz had begun to take on pupils of his own,
the result of a minor flaw in the fabric of the boy-scout idyll on the
Wilhelminenberg. From time to time there would be a sharp disagreement between Koenig and one of his students, and the student would
be summarily ejected: the honour of being the first to go is claimed
by Wolfgang Schleidt.

Born in 1927, Schleidt was old enough to have been sent to fight
against the Russians, but he had been shot and hurt just enough to
remove him from the worst of the war. The young man had then to
find an occupation suited to his temporary disability, and at the
Wilhelminenberg his interests in zoology and physiology were confirmed. Schleidt's conflict with Koenig came to a head following an
argument over the housing and safety of the Wilhelminenberg's
birds from animal predators. Failing in his demand for greater protection for the birds, Schleidt had provocatively placed the first inevitable
victim on Koenig's doorstep. The scene that followed would be familiar
to any ethologist studying primate behaviour: Schleidt's obvious
challenge to the group leader led to his immediate exclusion. It was
only by the good offices of Ilse Gilles that he was able to return to
hear Lorenz's open-air lectures and then to visit him at his home.

Lorenz listened sympathetically to Schleidt's request to come and
work at Altenberg, and to his explanation of the break with Koenig.
But he also wanted to hear the other side of the story and told the
young man to come back in a week. When Schleidt duly reappeared,
he found himself before a changed Lorenz who summarily dismissed
him and refused to listen to any further plea.

Soon Prechtl was thrown out of the Wilhelminenberg too, and he
and Ilse, now married, followed Schleidt in making application to
work at Altenberg, but more successfully. Within the year, Ilse inter-
ceded again for Schleidt and this time he was accepted—the pattern
of departure from the Wilhelminenberg being by then apparent.
Schleidt was to pioneer one of ethology's component fields, the study
of ultrasonic communication. For a man who was to work closely
with Lorenz this was a good choice: their interests did not overlap to
such an extent that Schleidt would be inhibited by the need to avoid
trampling over his master's territory. In addition, his ethological
interests were to some extent 'camouflaged' by his work on animal
physiology. All of this does credit not only to Schleidt's qualities as

an investigator but also to his capacity to learn from his experience of working with a brilliant, dominant, but occasionally temperamental and easily wounded leader. In the end, the society of ethologists demonstrates as well as any that human associations can be a great deal more complex than the animal societies they study.

This was a happy period at Altenberg. There were excursions into the Danube forest with Lorenz, thoroughly enjoyable conversations, and plenty of work for the Prechtls and Schleidt while Lorenz got on with his writing. The manuscript that Konrad had started in Russia was mentioned in an article that he wrote for the British *New Statesman*, and at once an English publisher, Peter Wait of Methuen, wrote to enquire after it. He was told it would become part of a standard work on ethology which was planned as four books of about three hundred pages each. Konrad had a publisher for the original German version, but no one in England as yet. That major work never materialised – although the prisoner-of-war work did eventually contribute towards a book that was published in German a quarter of a century later – and in English (as *Behind the Mirror*) not until 1977. A smaller and apparently far more trivial work came first.

Soon after his return home, while there were still no signs of a job, Lorenz had gone in to Vienna to give a lecture, after which he accepted a dinner invitation from a former student, Professor Sylvia Klimpfinger. Their host was her old schoolfriend, Dr Gerda Borotha, now a publisher and printer. The dinner conversation turned on children's books, and in particular a book about a bee that Lorenz pronounced stupid and misleading. All right then, someone asked, if others did it so badly, why did he not write a children's book himself? And Gerda Borotha said at once that she would publish it. So Lorenz went away and put together *King Solomon's Ring*.

The delivery of the latest instalment of the book became a regular evening entertainment for the young researchers with whom he shared his house. It was published in German as *Er redete mit dem Vieh, den Vögeln und den Fischen*, and put into English by Marjorie Kerr Wilson (later Marjorie Latzke), a friend and neighbour of the Lorenzes.

And so, in place of the vast treatise, this gentle 'children's book' came to England, and Peter Wait set to work to polish its English style the better to reflect the man who wrote it. An English ornithological adviser firmly recommended the deletion of references to jackdaws and other birds being purchased in shops and kept as pets; but that quibble was eventually banished to a footnote on the law in

Britain. The selection of an English title, too, was a problem: arguably the worst suggestion was 'At Home With Animals'. The one finally chosen also has its dash of confusion, for readers of Kipling's *Just So Story* of 'The Butterfly That Stamped' will find that talking to animals was a trivial matter for the legendary Solomon—his powerful Ring was needed only for much greater magickings.

Much of the experience of animals described by Lorenz is that of pre-war Altenberg, but the book was embellished by its post-war rounding out. A poetic distance is combined with the immediacy of the original observation, and the freshness of youth united with the nostalgia of the man who is searching once again for his roots. In place of the missing geese, the jackdaws, the deep comfort of his return, remind him more happily of his earliest years with animals. So it was that Tchock and his successors occupy the centre of his attention; in a book that purges the discontinuities of war, he repays his debt to their territorial loyalty.

The book carried his name to delighted readers around the world, its author presenting himself in his role as dilettante enthusiast rather than the serious scientist that overlaid this other aspect of him. His wife saw a danger in that far more clearly than the good that the book would do. But she later admitted, 'I was wrong. . . . I really thought it would spoil his scientific reputation by writing *Just So* stories. But it was the other way round. Very famous people came and said "Now I know what you want; now I know what you do".'

Among them was Professor Karl von Frisch, Director of the Zoological Institute at Munich since 1925, a leader in the study of the sensory capacity of animals and one of the most influential zoologists in Germany. According to Lorenz, von Frisch until then had hardly looked at a single ethological paper, but after reading *King Solomon's Ring*, he at once 'crammed' ethology and from that point onward made it his business to know everything about the science that Lorenz and Tinbergen had built up. 'And that alone', he says, 'is justification for having written that silly little story book.'

Perhaps it was an anecdote written about him that drew the senior professor's attention to Lorenz's 'story book'. Von Frisch had a parakeet which the cautious and methodical scientist would allow to fly free only after he had seen it clear its bowels, 'so that for the next ten minutes his well-kept furniture was not endangered'. As a result, the bird was quickly conditioned to produce whatever it could by way of a dropping whenever von Frisch came anywhere near him.

But whatever it was that brought him to the book, it placed him at once as a leader in the same discipline as Lorenz and Tinbergen, with whom he was later to share the Nobel Prize. He had already cracked

Niko Tinbergen K. v. Frisch   Konrad Lorenz

the 'code' of the bees' dancing 'language', and he was busy at that very time with experiments that showed the insects' use of polarised light from the blue sky to compute the direction of the sun. As with the other two men, much of the 'qualifying' Prize work had already been done, whether it be called ethology or simply zoology.

It was largely through the influence of Tinbergen that England in the post-war years became the second main centre for widespread ethological study. As a new and lively development in biology, this soon became a subject for excited discussion and then further investigation by scientists who might otherwise have trodden the routine paths of traditional zoology. As far as Britain was concerned, Lorenz and Tinbergen together were the joint founders of the science; of the two it was Tinbergen who, in 1951, was first to produce a book on *The Study of Instinct*. After 1952, this went a long time without further printings, for the ideas of Lorenz and Tinbergen were sharply challenged in the mid-fifties. Other British centres have since risen to vie with Oxford—notably Cambridge, with the already well-established Thorpe and then the powerful challenges of Robert Hinde's critical analyses of Lorenz's work, and his penetrating studies of individual development. All this has produced an English-speaking school of ethology to rival the German-speaking original, with Tinbergen, transplanted from Holland, the main link between the two.

Soon after finishing his first popular book, Lorenz set to work on a second, *So kam der Mensch auf den Hund,* published in English some years later as *Man Meets Dog.* Full of commonsense and wit, it also developed a theory already familiar in outline to readers of *King Solomon's Ring.* From the behaviour of the domestic dog Lorenz postulated that its principal forebear must be the golden jackal. The theory has the virtue that the behaviour of the jackal would have to be modified little compared with that of the other candidate forebear, the wolf. Lorenz frankly admits that 'we do not know for certain that it was exclusively the golden jackal that attached itself to man. . . .

However, it is quite certain that the northern wolf is not the ancestor of our domestic dogs as was formerly believed'. Exceptions to this, which by their very contrast in behaviour (as well as physical type) strengthen him in his conclusion, are certain northern breeds such as

Eskimo dogs and chows. Lorenz states his belief that man was already accompanied by jackal dogs when he reached the Arctic Circle and came in contact with the Arctic wolf, which he cross-bred with his already domesticated animals. Lorenz then proceeds to talk of 'jackal' and 'lupus' dogs and mixtures of the two among our domestic animals, expressing a strong personal preference for the dog that is biassed far enough towards the lupus (wolf) characteristic for it to behave like a pack dog that can be imprinted to accept his master as pack leader. In all this, Lorenz's descriptions of behaviour cannot be faulted; and his desire to breed dogs for greater intelligence (which implies a move in the direction of the wolf) is entirely reasonable. But his 'hunch' on the origins has since been strongly disputed.

Regrettably, this has become one of those areas of scientific controversy in which the leaders have little to say in favour of the work of the other principals in the field, a situation which is unfair to all of them and does credit to none. They include in this case Professor Wolfgang Herre of Kiel, and Erik Zimen, who has worked with Herre and also had associations with Lorenz in Bavaria. The common ground between them all is that dogs can, in appropriate circumstances, breed with both jackals and wolves; the question is one of domestication. Lorenz notes that dogs that have turned wild have been known to mate with jackals. Zimen has crossed a wolf with a poodle, and the result has been a hybrid in both appearance and behaviour.

Herre's main contribution has been in the comparative anatomy of the brains of dogs and their wild ancestors. As in some other domestic animals the brain of a big dog is thirty per cent lighter than that of a wolf of similar size; but when comparisons are made between a poodle and a jackal which are of comparable body weight, the dog's brain is found to be bigger than that of the jackal. No other animal is known where domestication actually increases brain size, while the loss of thirty per cent over many generations is common. In some basal areas of the brain the dog's loss is as great as seventy per cent, corresponding to drastic losses in sight, smell and hearing; but even this is in brains that are still, overall, larger than that of a comparable jackal. On this evidence it is reasonable to suggest that the ancestor of the dog must have been the wolf, and that the jackal can have played little part.

Zimen takes this further. He and his wife Dagmar have bred wolves in captivity and studied them using Lorenz's method; first at Kiel and more recently in the wilder surroundings of the Bavarian Alps. Wolves and poodles, Zimen notes, are both pack animals while jackals are solitary or live in pairs. In their threat behaviour, both dogs and wolves draw back their lips and snarl through closed teeth; the jackal puts his head lower, with his back bent up and the mouth opened. Physiological examination of the brain shows that specific areas (such as the visual cortex) are similar in the wolf and poodle but rather different in the jackal. Young wolf puppies show strong but undirected aggressive behaviour to their litter mates, then between four and six months this diminished to virtually nothing, but increases again between the ages of one and two years. The dog behaves like a very young wolf cub held back to the stage where there is simple, undirected aggression. This retardation is consistent with the fact that dogs lack the complex social structure of the wolf—in particular their mating behaviour whereby in the winter the two leading members of the pack are selected to mate while the others stand by, with the whole pack bringing up the resultant litter. Dogs have also lost the excellent visual capacity and some of the hearing power of the wolf. At close quarters wolves communicate with each other by visual signals, using their bodies; over long distances and particularly in the winter, they maintain contact by howling. Dogs depend much more on their sense of smell; visual signals are still used, but to a degree that corresponds with the differences in the brain.

Zimen finds that wolves can be tamed by taking them from their parents while their eyes are still closed (at less than two weeks) and imprinting them in the normal way; they can even be persuaded to pull a sledge. But he believes that the domestication process would take several thousand years of breeding by man.

The evidence looks strong that Herre and Zimen are right, and that Lorenz's hunch this time was wrong; so how do Lorenz's friends react to this evidence? Koenig takes the view that Lorenz could still be right—if the jackal forebears of the dog are not those we see today. Others have said it would do Lorenz little credit to give much prominence to his theory on dogs, and yet it is largely the methods that he has pioneered which have been used to support the argument against him. If he is wrong in this case, it actually strengthens his earlier argument on the general effects of domestication, to which man's best friends appeared as embarrassing exceptions. But whether right or wrong, with the ideas of which he wrote in *Man Meets Dog*, Lorenz has played his usual role of stimulus to others.

Chapter 9

The Scientific Exile

While busy on his popular books, Lorenz had neither the job nor the academic status that he felt to be his due. He was looking for something more permanent and with better facilities than could be provided by his house and garden at Altenberg, but had discovered that there was nothing for him in his home country.

Those post-war years were a time when the various military administrations laid extraordinary and irrational restrictions on German and Austrian scientists. One such was on equipment, particularly crucial items that could conceivably be used as weapon components. For example, oscilloscopes were all but outlawed (since they could conceivably be used in the development of guided missiles and radar) and special permission was required for their use in physiological research. Another restriction that affected Lorenz more directly was on travel: in the summer of 1950 special documents were required for his journey to Wilhelmshaven to attend a small informal conference that had been set up by von Holst. Because of sickness, von Holst had been unable to attend the Cambridge meeting of the previous year, although he had sent a paper. Now he had called together a number of students of behaviour to meet with the men who were investigating its underlying structures. Thorpe drove through Holland, picking up Tinbergen for his first post-war visit to Germany. On home territory were Kramer and Koehler, and von Holst brought in the physiologists B. F. Hassenstein and Horst Mittelstaedt. Hassenstein was studying the way in which the elements of insects' compound eyes interacted to register movement, while Mittelstaedt had been working with von Holst to show that movements in response to light were essentially independent of 'spontaneous' locomotion, and also, more generally, was investigating the effects of sensory feedback in cases where, for example, an animal can hear its own vocalisation.

Von Holst set the ground rules for the meeting: no proceedings were to be published, and only 'half-baked' ideas should be discussed.

The meeting was so successful that another one like it was planned for two years later, and this in turn has been followed by others at similar intervals. As the number of investigators has grown, however, these international ethological conferences have necessarily become major events of a more formal nature than that first meeting at Wilhelmshaven.

On his way home, Lorenz visited Göttingen and made arrangements for a post-doctoral position for Schleidt, little knowing that the situation for both of them would change soon and dramatically with a new offer. Kramer and von Holst, both Max Planck departmental heads at Wilhelmshaven, demonstrated that between them they had sufficient strength to push through an intermediate project which would provide Lorenz with a field station in Germany. The first stage of their plan depended on a scientific comradeship that had grown up between von Holst and an extraordinary German baron called Gisbert Friedrich Christian von Romberg, who lived in a castle in the Westphalian countryside to the north of the industrial Ruhr.

The Rombergs had a reputation for eccentricity: a nineteenth-century member of the family was a practical joker whose merry pranks were collected in a book '*Der Tolle Bomberg*' (a weak German pun on 'mad bomber'.) Early in the Industrial Revolution the family had begun to exploit the coal beneath their lands in the Ruhr; and among the pioneers of deep-mining was a Romberg whose grandiose inventions include that practical money-spinner, the lifting cage and winding gear whose towers and wheels are still to be seen over coal pits in many parts of the world. The resulting wealth of his descendant was applied to science rather than technology, for Gisbert von Romberg dabbled with dilettante near-genius in physiology, exploring the eye tremor of miners. From this he branched out in the general study of eye movement, where his interest overlapped that of von Holst.

The Romberg property at Buldern bei Dülmen, some twelve miles south-west of Münster, was centred on a seventeenth-century mansion that had been expanded in Napoleonic times to become a chateau surrounded by 'English gardens'—the German term for open, landscaped grounds. The buildings were already partly occupied by departments of Münster University which had lost their home in air-raids. In December 1950, Lorenz and his assistant Schleidt were provisionally installed in the old servants' quarters while more spacious accommodation was being prepared. From Vienna came the Eibl-Eibesfeldts and the Prechtls, and a skittle alley was pressed into service as accommodation for the four of them and their animals. (Otto and Lili Koenig, by now firmly established at their Wilhelminenberg Biological Station, stayed on in Vienna, and, indeed, were still there many years later to welcome Lorenz back to Austria, playing then a role similar to that of Gretl Lorenz, who had kept Altenberg for Konrad's return from Russia.)

Lorenz could not have full 'Max Planck Institute' status at Buldern because permission at that time, not readily granted anyway, would have been required from both civil and military governments. His establishment as head of a new sub-department within an existing Max Planck Institute—the next stage of the plan—was a purely internal matter, but even that took all the persuasive skills of Lorenz's two friends against opponents who did not regard his work as 'hard' science. To Buldern came the news of battles and skirmishes that shifted prospects this way and that until at last a victory was won; Lorenz got the go-ahead—and the promise of a slender budget—on the first day of April, 1951.

After the makeshift conditions of the years before, any budget at all was riches enough to provide a foundation for the expansion that was to follow. Lorenz's old friend Otto Koehler sent some of the first post-doctoral students to Buldern, including Margret Zimmer, who was to be Schleidt's first wife; and a striking, dark-haired girl of part-Polynesian ancestry, Beatrice Oehlert, who would marry Lorenz's son Thomas. For several years, Thomas sought to make his way as a painter, but had to return to physics a few years later when art denied him and his family an adequate living.

A single room doubled as Lorenz's bedroom and study, and also housed some seventy fish, a number of ducks and geese, and thirty or forty of Ilse Prechtl's songbirds. When his daughter Dagmar came to stay there was no other space available so she camped in the same room, which served in addition as a breakfast room for the small community and for informal lectures by Lorenz. One day, the main aquarium burst and flooded the place to a depth of two or three inches, offering everybody the sort of experience that is funny in retrospect. Fortunately, that room was only temporary accommodation; as soon as they could, Lorenz and his family moved into an old mill that was part of the property. The others went into the older wing of the castle, while a greenhouse was pressed into service as an aquarium, and the skittle alley became an animal-holding facility.

In due course, the nearby University of Münster, already associated with Buldern through the departments that were housed there, took Lorenz under its academic wing and found him an Honorary Professorship, so the important matter of status was satisfactorily settled. The day came for a visit by the President of the Max Planck Gesellschaft, the celebrated nuclear physicist Otto Hahn himself. For this special occasion, Lorenz found a suit and tie. Showing his distinguished visitor around the establishment, he talked with his customary enthusiasm until Hahn stopped him dead by asking, 'Tell me, are you *really* naive, or are you only simulating?'

Lorenz's more usual peasant costume was topped by a woollen hat, and from the angle at which he wore it the assistants could tell his mood. Set well back on his head it signalled high spirits, but as its band came down lower over his brow, correspondingly greater caution was required in approaching him. When clamped down tight above his eyes, it told of a very sore head indeed — which was useful information, for his mood could change quickly. Prechtl recalls the morning when Schleidt went to him with the news that the aerating pump for the aquaria had broken down and the fish were already showing a tendency to float upside down. Lorenz pulled his woollen hat lower, growled, and retired to rest. Schleidt disappeared to Münster to have the pump repaired and when it was functioning again, went to report that the fish were reprieved. 'Splendid, splendid!' cried Lorenz, and rose to tackle the work of the day, his woollen bonnet once again at full mast.

Their host, Romberg, reserved for himself the grander rooms of the castle, where he maintained his own scientific enthusiasm in style. The sixty-three-year-old baron was a crank, a mysogynist and an alcoholic who regularly got drunk for days on end. Lorenz's assistants took turns at keeping him company during his recovery from these bouts.

While Schleidt was building up Buldern's population of turkeys, Eibl-Eibesfeldt set to work with small mammals such as stoats, hamsters and squirrels, and was soon winning high regard for his results. In squirrels, Eibl worked out the nature of their nut-opening technique. Faced with a nut, these animals know without previous demonstration that they should gnaw at it, but often did so initially in the wrong direction, across the grain. After a few nuts they would work out that if they gnawed along the grain they could get it open quicker. Eibl's other work included a re-investigation of nest-building by rats. He had read a paper which said this behaviour all had to be learned, and that rats raised without suitable materials to hand could not do it; but Eibl was able to show that the elements of the nest-building operation were already there although the sequence itself had to be learned, resulting in a gradual improvement in performance. The earlier negative result could be explained by the inhibition the rats suffered if put in a new, special cage for experimental observations.

Among the fish, the cichlids continued to offer new insights. Beatrice Oehlert chose species where the male and female were identical in appearance in order to discover how they would work out whether a companion newly placed in a tank with its territorial 'owner' was male or female. She discovered that this was established not by what they did but by what they did not do—a subtle distinction, but one that demands a great deal more of the observer than simple watching and cataloguing. She noted that male and female cichlids combine their innate behaviour patterns differently: the male is capable of interlacing aggressive behaviour towards his companion with displays of enticing sexuality, while the female can intermingle flight and sexuality; the reverse combinations do not occur in either sex. In another experiment she placed mated pairs of cichlids in a single tank but separated them by a pane of glass. The aggressive males did their best to fight each other, but their mutual threats gradually decreased as a green algal layer grew across the intervening pane. Then they turned inward, each male to his own mate; the amount of redirected aggression correlated with the rate of growth of the algae on the glass.

Tinbergen sent students from Oxford, including Rita White who worked with pigeons, and for whom he considered it would be good experience not only to join Lorenz but also to practise a language that was valuable to any ethologist of that time; many of the early papers were available only in German. Another student who had worked with Tinbergen at Leiden was Uli Weidmann, a German-speaking Swiss. Buldern seems to have been a favourable place for pair-formation among the ethologists as well as their animals, for before they left, Rita and Uli were married—which meant that she found her German specially useful and he came to work in England where he is now at the University of Leicester.

While at Leiden before the first post-war meeting between Lorenz and Tinbergen, Weidmann had heard not only of Lorenz's work but also of his former political sympathies, and it had been with some apprehension that he had heard Lorenz would be calling at Leiden on his way back from England in 1949. As it transpired he need never have worried, for the visitor was not the authoritarian he had expected, but an easy-going man of great personal charm and magnetism who talked to him not about animals but for hours on end about the philosophy of the perception of self. Weidmann also saw (then and on other occasions) Lorenz and Tinbergen reunited after a period apart, each time producing a firework display of intellectual excitement, with Lorenz playing the visionary and Tinbergen always probing, redirecting, testing and asking 'are we clear enough on this?' as they developed their ideas on the nature of instinct.

When he reached Buldern a short time after the research station had begun, Weidmann had no money to support himself, but he was able to earn a little as the Institute's duck warden, sleeping and spending most of his time in a small wooden shed on the edge of the artificial lake. For his Ph.D. thesis, he had unrivalled opportunities to observe the fixed action patterns of the mallard duck, and was even able to disprove an earlier hunch of Lorenz, that three of its display movements were of equal significance, and that chance would determine which was chosen by the animal on any particular occasion. By that time a great deal had been written on display, but none of it was quantitative. Using statistical methods, Weidmann demonstrated that chance would have produced different results from those he saw.

At nearby Münster University, a biology student named Wolfgang Wickler had been working on plant organelles for his Ph.D. thesis, but on hearing Lorenz lecture he switched abruptly from botany to zoology, and looked for an animal to study. With myxomatosis spreading across Europe, work on rabbits — the first suggestion — seemed ill-timed. But then, from a trip to the Italian lakes Lorenz brought back a chianti bottle containing specimens of blenny fish, curious animals with no swim-bladder that hobble about on the water's bottom. Yet these are no dull, stunted creatures; despite their disability (or because of it?) they manage remarkably well. So here was a mystery for Wickler to tackle: he set himself to compare them with other fish which had independently evolved the apparatus to live as bottom-dwellers. But more interesting to Wickler than any particular species was the scientific method itself, the technique by which it could be studied. What had attracted him from the first was not the blind wish to do as Lorenz did, but the wide-open opportunity to follow in the path that Lorenz had blazed and to organise it by the invention of his own more precise scientific methods: with these,

117

Wickler was determined to make his name—as he did, eventually becoming a successor to his teacher.

Lorenz's new studies of geese were just beginning. A World Health Organisation meeting in Geneva, in January 1953, was attended by Frank Fremont-Smith, the then Director of the Josiah Macy Jr. Foundation of New York. As a result of this, the post-war goose studies were supported by American funds. On a visit to Buldern, Fremont-Smith talked of a parallel investigation that would range Lorenz's growing gosling population alongside a cohort of American children in a comparative study of early development. A 'longitudinal' study of forty thousand children did in fact start in America in the late 1950s, but typical of the hundred and sixty variables measured in that incomparably larger investigation were such peculiarly human attributes as 'socioeconomic factors' and I.Q. (which predicts school performance rather than the intelligence its name implies). The inheritors of this American National Study say that more interesting and properly comparative factors might have been chosen today, but it would still be most unlikely that any such programme would be undertaken without a great deal of measurement. The Buldern goose study – which by Lorenz's methods would have little or no measurement at all—went ahead on the assumption that there would be observations on such social matters as the time the young spent with their families, the rank ('pecking') order of siblings, and so on: ten years of parallel work, it was thought, should be sufficient. Another Ph.D. student, Helga Fischer, was brought in from Münster University to specialise in their care and study. For her, this turned out to be a longer job than anyone imagined: she was still working with the geese two institutes and over twenty years later. But no scientific comparison with human development was ever made.

Under Lorenz's loose rein formal organisation was minimal, and his small community enjoyed a brief period of busy tranquillity and growth, until their patient animal observations were sharply interrupted by the death of Baron von Romberg in the summer of 1952. It soon became apparent that the heirs to the estate were not interested in any science concerning dogs, geese or men: they were more concerned with what could be done to realise the value of the property. Lawyers moved in and Lorenz and company were requested to find themselves another home. Perhaps the Romberg heirs were more forbearing than this story makes them appear, and it is also possible that the prestige of the Max Planck Gesellschaft itself played a part in events, but whatever the cause, the scientists and their geese actually took several years to move, and the work of the younger men went on.

But at last the time was ripe for Kramer's three-department establishment. Lorenz and von Holst, with the promise of solid backing from the Max Planck Society, now began a methodical survey for a place to set up the new and greater institute that would finally bring them together. They preferred Southern Germany, and systematically combed the countryside from the Swiss border to Salzburg. Their criteria included suitability for geese, a rural setting that was not too far from a good university, and peripheral matters such as satisfactory game laws.

About twenty miles south-west of Munich (which could provide the required professorship), they found their ideal in the rolling country-side of the Lake District of Bavaria, between the Ammersee and the Starnbergersee. Here was the tiny pear-shaped Ess See, surrounded by grassland and woods, isolated from the neighbouring villages, and far enough away from the bigger lakes for the geese not to be tempted to stray. The fields on its eastern shore—marked 'Seewiesen' (lake meadows) on a local survey map—were available, as they had been in the hands of the local State Government since 1938 when they had been taken from the monks of the nearby Monastery of Andechs for non-payment of taxes to the Hitler Government. Tax evasion is against the law under any government, and somehow the property had never been returned.

Any meeting between von Holst and Lorenz was liable to result in a vigorous argument over the fine points of their science or ethics. Von Holst had a strong personality, and with it, was as arrogant as only a Baltic Baron can be, while Lorenz gave as good as he got. Between them, Kramer excelled as a patient moderator. The conflict never seemed to worry the two protagonists, although Gretl did get upset, and that indirectly affected the two men. The upshot was unexpected: it was Gustav Kramer, its initiator twenty years before, who eventually pulled out of the Seewiesen project, thereby removing one leg of the tripod on which it rested. He would still be part of the Institute, but his bird orientation work would be conducted far away at Tübingen in the south-west of Germany.

On the Seewiesen shore, in 1955, the main research centre began to take shape. First there was a fence around part of the property, erected in the hope of keeping foxes out, and then on the east bank a large cage and a small house, the Gänsehaus, to provide urgently needed security for the precious geese—who stood under imminent threat of death from the shotgun of the new Baron at Buldern, whose inherited guests had long outstayed his patience. The new establishment was called the 'Max-Planck-Institut für Verhaltensphysiologie' (the physiology of behaviour), its very name reflecting the 'hard' science image that seems to have been necessary to counter the unfashionable 'softness' of Lorenz's method. And unlike Peter Scott's wildfowl, Lorenz's geese were to remain protected from the gaze of outsiders.

The Schleidts were the first to move in, and they acted as caretakers, guards and resident representatives of the incipient Institute – struggling with and mediating between architects, various contractors, and the distant directors; von Holst was still at Wilhelmshaven and Lorenz was holding the rear at Buldern where the situation had become tense. He finally moved at the end of 1956. On the lake itself, Lorenz could expand his goose studies to embrace some eight separate species under the foster-motherly care of Helga Fischer and her assistants. The idea, as at Altenberg before the war, was to spread his comparative method over a wide range of creatures in the expectation that the similarities and differences should be productive.

By the time it was finished and everybody had arrived, the new Institute was very much larger than the original, intimate Buldern group and a little overwhelming in its composition: it had something of an ivory tower atmosphere, with a succession of post-doctoral scientists moving in from England, America and Holland. Each Wednesday afternoon there would be a colloquium which brought all the various interests together—and here the points on which Konrad Lorenz and Erich von Holst failed to see eye to eye could be allowed full rein. Their lively arguments did much to forge the spirit of the new Institute. In Lorenz's own eyes, his work before the war was the best; for Schleidt and Prechtl the golden years had been at Altenberg after Lorenz's return from Russia; for those who had joined Lorenz at Buldern the heroic times were then: for each, the periods that followed could never match those that had gone before. Now, yet another generation had arrived to designate the early years at Seewiesen as the greatest. These newcomers included some whose introduction to ethology had come from reading *King Solomon's Ring*.

Sverre Sjölander from Stockholm was one of many now in universities all over the world who regard themselves as Lorenz's 'ethological children'. As a schoolboy, Sjölander obtained Lorenz's address and wrote to ask if he could visit and help in some way. At that time there was no students' accommodation and no canteen, so the unpaid helper was simply invited into the Lorenz house at Seewiesen during his school holidays, and Gretl provided his meals, which is typical of how it was: Konrad handled all relationships within his department at a personal level, and Gretl was deeply involved too. Buldern, and then Seewiesen, were like Altenberg, but with more space and progressively better facilities for the scientific studies. At its best, the system—which was totally natural to the Lorenzes—worked superbly. But the reverse of this coin showed 'out of sight out of mind'. If Lorenz or a member of his Department was away for a while, matters which depended upon the reminder of frequent contact could hang fire, which was worrying, especially for any young man seeking preferment.

When Lorenz did go away for a while, assistants and staff would assemble inside the gates to wave him off, in half-family, half-feudal spirit. On one occasion when the scene was set for this little ritual, someone came running out with a forgotten item of luggage—a chamberpot belonging to one of Agnes' children. Konrad took it, raised it solemnly above his head as a crown, and drove slowly away from Seewiesen as a mock feudal lord.

Each year Konrad and Gretl gave 'tremendous' Christmas parties; at other times he would swim in the moonlight with members of his staff. Most of his younger associates enjoyed this informality, and his occasional displays of clowning; but a few, inevitably, would have felt more comfortable if their teacher had displayed a little more of the traditional protocol to be expected of the great professor.

The relative tranquillity of Seewiesen was short-lived, for all too soon the young Institute was hit by tragedies that were to deprive it of the two men who had joined with Lorenz to make it possible. Their loss was to be followed by changes that would redirect much of Seewiesen's science away from the course Lorenz favoured. The first to go was Gustav Kramer, at the age of forty-nine. A site had already been bought at Tübingen and the foundations of his laboratory laid when news came that he had been killed while mountain-climbing. That was in April 1959. Then, in 1961, the ailing von Holst relinquished the directorship and in May of the following year died in hospital at nearby Herrsching. Two new department heads, Professors Schneider (insect olfaction) and Mittelstaedt (cybernetics) succeeded Lorenz's lost comrades, and Lorenz himself was confirmed as Director.

In retrospect, this has been seen as the end of the 'great' years—great by whichever generation of ethologists they be defined—for the unity born of vigorous controversy began to wither after the loss of Holst. Yet Seewiesen continued to grow: there were more buildings, more assistants, and more students, but as it became bigger the Institute became more impersonal. There was still an intellectual excitement for the new students, but this came more often now from Wickler or Schleidt. For Director Lorenz, the old personal touch was no longer possible; he must become part of the administrative machine, to be seen by appointment, with much of his time already commanded by his responsibilities as a 'big' international figure. For some of his former pupils this meant that he could be seen rarely, if at all; a few, perhaps with justice, felt they had been forgotten.

For a while, Gretl continued to help her husband as before, but this was not to the liking of the Max Planck Gesellschaft with whom she had to deal on matters concerning arrangements at the Institute; their attitude was, 'We will deal with the Professor, not his wife'. At this, Gretl was hurt and withdrew somewhat, which led in turn to a decline in her husband's enthusiasm for his own role as Director. A great deal of the early, easy atmosphere was lost, although the scientific work of the Institute continued, much of it with notable success.

Before he died, Kramer had recruited Jürgen Aschoff, who had wanted to switch from cardio-vascular research at the Medical Institute at Heidelberg to the then almost unknown study of biological rhythms. Aschoff came to Erling-Andechs, a few miles from Seewiesen, where a villa had been given to the Max Planck Society. Since then the Erling group has provided a notable success for the Institute, and Aschoff himself has become a world leader in the study of biological 'clocks', the still mysterious processes that operate in plant and animal systems to produce the daily, monthly and other rhythms that so deeply affect all our lives. Wolfgang Schleidt was an early subject of one of the isolation experiments on this project. The procedure removed all time references from the experimental subject, letting him eat, sleep and choose light or dark as he felt the need. In those days, nobody knew how people might react to such experiences, so great care was taken with the 'volunteers'. Schleidt certainly came to no harm, and in fact took the opportunity to get on with some uninterrupted writing. Experiments like this have shown that man seems to have a daily cycle that, when allowed to run free of all constraints, lasts a little over twenty-five hours, but which is regulated in each successive day of normal life to twenty-four hours in response to cues from the sun and the regular events of any human society. 'Chronobiology' has become a mini-science in its own right, and Lorenz is proud of the part that he and his friends played in helping to get it started. It is a direct and logical outcome of the synthesis of the interests that brought them together: the investigation of biological spontaneity, and the way in which the centres governing instinctive patterns must be primed by internally produced impulses, at least partly under the guidance of some inbuilt clock.

For many years, an outstation at Schloss Möggingen near Radolfzell on Lake Constance came under Lorenz's direction. With its investigations into the mystery of bird migration, this closed the circle of the original joint concerns of Kramer, von Holst and Lorenz. In recent years, its management has been transferred to Aschoff, whose relationship with Lorenz has fortunately never been confused by the overlap of common territory.

One line of research at Seewiesen itself was dictated by queries that behavioural scientists in other countries began to raise around Lorenz's earlier work on imprinting. For all the length at which he had discussed this in his 'Companion' paper of 1935, it was still more a method that he used than a major study in its own right; he had never really examined it experimentally in any detail. So when he felt his ideas on imprinting to be attacked in these post-war years, he was unable to point to any solid or acceptable scientific backing for what was undeniably an important cornerstone of much of his work and conclusions; such an investigation was both vital and long overdue. Perhaps he took the challenge more directly and personally than was really necessary, but perhaps he also felt that without new support there was a danger that the credibility of the whole structure of scientific theories he had built up might be undermined. Whatever the reason, it was now a matter of urgency for Friedrich Schutz to take a closer look at some of the assumptions made by Lorenz so many years before.

Schutz was a former student of von Frisch who turned his attention from the 'alarm substances' of fish to Lorenz's geese. Ethological research in Britain was also well under way by this time, and it was one study in particular that presented serious problems for the goose workers at Seewiesen. A young student of domestic chickens at Cambridge, Pat Bateson, re-examined some of Lorenz's work on imprinting and found that what had been reported from Altenberg did not seem to work in the same way with his own birds in the more rigorous conditions of a laboratory study. Was something wrong with Lorenz's classical description?

'With some justice', as Bateson has since agreed, Lorenz objected vigorously when the evidence of non-imprinting in one species was used to refute what he had seen and described in another. As Bateson discovered and Lorenz already knew, animals vary considerably in their response. Geese are less affected than jackdaws in their subsequent sexual behaviour, while some animals (including some birds) will produce the following-reaction to order, but little else: Bateson's laboratory chicks seemed to be a case in point. If the essence of science is repeatability, this does not necessarily mean that a particular type of stimulus must produce the same result in every species, though the lack of such an extension is bound to limit the more general conclusions that might be drawn.

Another entirely reasonable aim of scientific method is to reduce the number of variables in a given situation, the better to assess each individual influence. In a later case where it again appeared that some of Lorenz's results were unrepeatable, the answer was that in the laboratory situation the young birds were simply not being given the maternal evidence of love and affection that Lorenz's hatchlings automatically received in the more natural conditions of their imprinted upbringing. Pupils of Lorenz demonstrated that if a gosling was ignored except for its purely material needs, its adult behaviour was disturbed and abnormal, and a similar experiment carried out in America with human infants (one which in retrospect seems totally inhuman) showed that the same thing happens in mankind. Loving care in childhood is needed for mature social behaviour to develop. At this point, the analogy between goose and man is valid.

If we move forward in time to Seewiesen in the sixties, evidence of such reinvestigations lay all around. The most impressive sights to greet a visitor, and those which became much identified with Lorenz at that time, were the studies of the relationship between imprinting and the later sexual behaviour of some of the bird species first studied by Lorenz before the war, and others which his followers and assistants were still studying by his same methods.

In one example, mallard eggs were placed under a shelduck, so that when they hatched, the mallard ducklings would follow the shelduck mother. When they grew up, some of the mallard drakes sought to mate with sheldrakes, assuming incorrectly that these were suitable females (they completely ignored their own species). The female mallards, however, did not make the same mistake: the duck evidently has a sufficiently strong instinctive understanding that her mate should have a green head. Along with Bateson at Cambridge and some others, it fell to Friedrich Schutz to patiently unravel, over a period of fifteen years, some of the mysteries involved. At Seewiesen, the research had to be more analytical than was usual for Lorenz, while to extend their own understanding several experimenters in other places finally carried their studies out from the laboratory to more natural surroundings.

At the superficial level, the scene at Seewiesen was one of animal perversion rampant. The visitor could see a hand-reared muscovy drake courting Schutz and attempting to copulate with his out-stretched boot. In another example, a greylag goose had become paired with an imprinted mallard drake when the goose accepted a gander as well. The three lived together in moderately confused harmony, only occasionally quarrelling as they collectively reared their shared family. In some cases, cross-breeding that is not seen in the wild was possible here, and when this was done the behaviour of the offspring hybridised in the same way as their form mixed the physical characteristics of the parent species.

Once imprinted, behavioural perversions can remain for long periods unchanged: imprinting is not learning but the redirection of an already existing instinct. A pair of homosexual Carolina drakes (reared only in the company of males) persisted in courting each other for a decade, each behaving as the male and seeing the other as female. In most cases of cross-species imprinting, the differences between the sexes are no longer recognised, but Lorenz did have a mallard that could distinguish correctly between the sexes of crested chards upon which it was imprinted—and that it should have been able to do so is more surprising than if it had not.

Schutz feels that some misunderstandings have risen over what is meant by 'learning' in the imprinting process. Learning to recognise a mother is entirely different from the trial-and-error process that is also called learning—when, for example, the gosling nibbles at different materials around him to discover which are good to eat and which are not. In another type of learning, the fledgling jackdaw is taught to recognise a common enemy by the hostile reaction of the older birds around it. If Lorenz has also used the term 'learning' in the context of imprinting, it is not surprising that his readers or listeners have not always understood clearly that what he is talking about is different again.

In the case of sexual imprinting, the consequences of something that happened to a duckling when it was perhaps a month or so cannot be recognised for many further months or for as much as a year later: the effect lies dormant. In extreme cases, sexual behaviour may appear normal for a year or so and then, when a normal partner is removed and a member of the imprinted species introduced, the behaviour reverts: the experience of the early weeks can have a stronger effect than that of some months or years later. This can be explained by supposing that the action is based on an instinct that is combined with early misinformation.

Schutz went on to differentiate clearly between the imprinting of the following reaction and the imprinting of later sexual behaviour—in Lorenz's work these had not been separated. In one of Schutz's experiments, a duckling was hatched and reared by a mother of another species for between five and ten days; a strong following reaction developed. After this initial period, the duckling was removed from its foster mother and reared in isolation, and during the next year it was offered the choice between its own species and that of its foster mother. The duckling chose its company from and mated within its own species—not that of the foster mother. Hence, the experience of the first days of life—the perfectly 'normal' following reaction—had had no effect on its subsequent sexual behaviour. In another part of the same experience a duckling was reared with a foster mother for thirty days or more, and then isolated until the following spring when it too was offered a choice of companions and partner. This second duckling did show altered sexual behaviour, which demonstrated to Schutz that the two forms of imprinting are not the same.

The findings of Schutz and others have vindicated Lorenz's statements on such matters as the stability of acquired sexual preference, although Bateson still disputes Lorenz's subsequent inferences. But that is all good scientific in-fighting, the debate that accompanies progress. Lorenz has stimulated research, whether or not the detail of the final conclusions is as he would prefer. In the event, his own work did not have to be dismantled and, in fact, gained strength in the process.

During his early years at Seewiesen, Lorenz had been writing what would become his most widely influential book, *On Aggression*. But that itself was only the second stage of an already rapidly escalating international controversy over the relative importance of instinct and learning—'nature and nurture'—in animal behaviour. While at Königsberg, Lorenz had already felt the need to defend his concept of instinct from those who might say that it denied the freedom and mystery of a God-given vital spark of life. But in the post-war years, as his ideas and influence began to cross the Atlantic, he met opposition (far stronger and of a directly opposite kind) from those who saw all behaviour as deriving from post-conceptional experience. These two challenges have occupied much of his thinking through the middle years of his working life. They are the contrasting concepts of Vitalism and Behaviourism. A world without 'isms' would be a great deal simpler.

Chapter 10

The Message in the Animal

The behaviour of all living creatures depends on the interplay of the equipment with which they are born and the events of the world around them. There has been much argument around this apparently simple idea, leading to impassioned controversy between 'nativists' and 'environmentalists' while others have attempted to withdraw from the battle in order to get some work done, probably thinking that the argument was more over the meaning of words than the ideas behind them. (If that is the case, the argument is in vain for the same work will be done in the end.)

The debate is complicated by the fact that 'all living creatures' includes man. There are many who would claim that man is so different from other animals that little that can be said about 'animals' has much relevance to his own behaviour. Even Lorenz recognises that this must, to some degree, be true — witness his approval of the statement 'all animal is in man, but not all man is animal' — but the degree of overlap is only the start of the problem. It is compounded by an accident of scientific history, that the examination of man's behaviour began earlier than the intensive study of the behaviour of animals. So it is hardly surprising if the objective study of man is still overlaid by subjective or philosophical overtones, for this is how such studies were conducted long before the scientific method was recognised, and psychology in its many forms has evolved from the philosophy that preceded it. Religious, political, ethical and aesthetic considerations are among the influences that bias man's observation of his own behaviour; indeed, it is Lorenz's contention that it is an important part of man's nature for this to be so; that it is there as part of our equipment for survival.

The application of ethology to man is the subject of later chapters. Here we shall be concerned primarily with animals, but in doing this we must always be aware that as people talk about other creatures, they may well be thinking simultaneously of how this could apply to man, which can affect what they say and how they say it. It is hardly surprising that there should be energetic attempts to mould the findings of animal behaviour into the existing models that have been established for man, or to reject them if they do not conform. Darwin himself felt the explosive force of this reaction. His *Origin of Species* concentrated most strongly on the form of animals and man, but it did not escape his critics that with the evolution of form must come the evolution of behaviour, and so his ideas were rejected by many. In their view there could be no such overlap between animal and man.

For long after Darwin it would have been more comfortable for scientists working on opposite sides of the fence if man and animal had as little in common as possible, so that while the principle might be accepted, not too much existing but otherwise conflicting science (let alone philosophy, politics and religion) need be thrown away. Darwin said quite explicitly that he most certainly intended to include behaviour along with structure as having been fashioned by evolution, and wrote a book to support this view. But the statement and the book were allowed to fade from general view; and the few scientists who remembered and whose work reflected Darwin's ideas on behaviour – men such as Charles Otis Whitman of Chicago – did not represent a mainstream of science.

The beauty and simplicity of Lorenz's position was that he was able to develop his early ideas in almost complete ignorance of the rigid structure to which a more formally disciplined scientist would have been expected to conform. He fitted into a separate and accepted tradition, that of the animal lovers and ornithologists who observed simply for the joy of observing, unbiased by the ideological arguments with which they were not involved; neither did they influence the course of such disputes. Lorenz's early development came almost directly under the influence of Darwin, and his path was not substantially changed by contact with Heinroth. So his work grew to a stage where he had a body of science of his own that was virtually independent of the mainstreams of philosophy and psychology that co-existed with it. Where he met ideas which conflicted with his own studies, he could simply reject them as being contrary to experience. But where he met a sympathetic scientist such as Craig (the pupil of Whitman), he maintained the contact and developed further in his own established direction. He interacted in a similar way with von Holst: here was a contemporary who had studied behaviour as a function of neurophysiology and was converging towards Lorenz's line; so this too was acceptable and helped to build up Lorenz's position. The third type of interaction that worked was that with Tinbergen, where objectives were closely matched but approached by different methods.

The picture of Lorenz and his field that begins to unfold is not that of a conventional scientist filling a previous void in human knowledge, but of a man who blithely continued to expand his range long after any widely recognised void had ceased to exist. To take an image from Lorenz's own science, ethology had overlapped deep into the territory of competing sciences long before there was any real contact between the opposing forces. The reason in one case has already been given: it was because the territories, in the physical sense of the places where the different ideas were held, were geographically far apart. Behaviourism, which emphasises the role of learning as a determinant of behaviour, was in the ascendant in America, while ethology, more concerned with instinct, was growing in Europe. The conflict when it did come was all the stronger for the amount of common scientific territory that was in dispute. For Lorenz, as for some American behaviourists like B. F. Skinner, the nature-nurture controversy, as it came to be called, is still very much alive, although for most biologists it is now receding into history. The heat went out of that debate with the formation of a third group, that called by Lorenz the 'English-speaking ethologists', and their truce with psychologists of various schools. In England and increasingly in America, this middle position may be a fruitful synthesis, but to Lorenz it is a political compromise. He is not only proclaiming that instinct and learning (as he understands them) are separate and distinct, but also rejecting the 'sell-out' of those who he had thought should be his friends.

The philosophical background that made Behaviourism acceptable in America was the idea of the equality of man that grew up in the French Revolution. Despite all the changes that have taken place in practical American politics, the ideals that were carried across the Atlantic nearly two centuries ago still shine freshly as though they were not contradicted by nine-tenths of the society that surrounds them today. For many, it is an article of faith that men are all born equal and that only inequality of opportunity together with the efforts that each individual makes on his own behalf gives each man his different place in society. This belief is not entirely consistent with Behaviourism, which is highly mechanistic. But to the true believer in equality, Lorenzian ethology is very much less acceptable when it

The Lorenz house built by Adolf in 1903 in a mixture of Austrian baroque and American megalomaniac styles. The lane leads up to the northern hill of the Wienerwald.

The mischievous boy Konrad at nine, and the poised High School student of eleven—he was then attending the Schottengymnasium in Vienna.

Adolf Lorenz on his 70th birthday. 21 April 1924 was the day of his retirement from the University of Vienna.

'Mother' Konrad feeds his imprinted goslings on a backwater of the Danube at Altenberg in 1937—during Tinbergen's visit.

Konrad with some of his jackdaw family in the fields of Altenberg in the 'thirties.

The morning ritual from the house to the Danube, and back again.

Part of the Danube forest near the Lorenz home — these are the marshy backwaters where Konrad and Gretl played as children — taken by Niko Tinbergen during his visit to Altenberg in 1937.

Gretl Lorenz in 1973, at the time of Konrad's return from Seewiesen to the family home at Altenberg.

Man meets dogs—at Seewiesen.

A rare experiment involving Lorenz: with microphone and miniature loud-speaker he makes the model mother emit motherly noises. It was later discovered that the shape of the model was immaterial. Essentially, only the movement and sound matter, although this varies from species to species.

Lorenz receives one of his many honorary degrees from the Chancellor of
Birmingham University, his old friend Sir Peter Scott, on 15 May 1974.

Konrad Lorenz and Niko Tinbergen
at one of the two-yearly ethological
conferences, in the early 1960s.

Lorenz with imprinted snow geese, swimming in the lake at Seewiesen — each morning they swam together in the Ess See.

Goose girls, surrogate mothers and observers of the goslings' behaviour, with their charges on the day of their arrival at Grünau in Almtal, Upper Austria — taking their midday rest.

Researcher Charlie Thomforde and his charges at Grünau—with the new
Institute behind—familiarizing them with their new surroundings in
Summer 1973.

Konrad anxiously tending a wounded
goose after the move from Seewiesen
to Grünau.

proposes instincts and pre-programmed individual capacities: this is all far too fatalistic.

As a school of psychology, Behaviourism dates from 1914 and the pronouncements of J. B. Watson, who made good sense in his attempt to turn the study of behaviour away from introspection and the contemplation of intangible and unmeasurable mental states and replace it with objective methods based on those of Pavlov. Watson was an early ethologist whose basic work on the tern (on the Eastern coast of the United States of America) is a classic in the field: he was perfectly aware that much animal behaviour is innate. In man, however, he held the innate component to be almost totally unimportant (except in extreme and abnormal circumstances). Many of his ideas can now only be described as 'nutty' ('thought exists only as sub-vocalised speech'), and he grossly oversimplified the conditioning factors of childhood (the only influences he thought important were loud noises, the fear of falling down, the restriction of movement, and caressing) to a degree that became embarrassing to his followers. But whatever his shortcomings, he did create a school of philosophy that survived for well over half a century and still has strong influence today.

According to Lorenz, the way to Behaviourism was opened up by reaction against an earlier philosophical approach to the study of behaviour which was called 'Vitalism'. This enshrined the idea that in addition to the physical processes of the brain there exists some vital spark—an external natural force—which in the final resort directs behaviour. As a concept, Vitalism has the advantage of being acceptable within a whole range of existing systems of ideas. It leaves a gap into which the religious man can drop a soul, or a philosopher conscious-ness; it readily allows for free will, and so for good and evil. It even gives the scientist room to manoeuvre, for so long as he leaves the vital gap, he can discover all he likes (or is capable of) about the processes of the brain, and not get into trouble. But there is all the difference in the world between a gap that exists because the scientist does not know the answer, and a gap that actually is the answer. In scientific terms, Vitalism leads nowhere: that which can explain anything at all explains nothing.

Lorenz grew up with what he describes as 'a healthy abhorrence' for such woolly thinking, especially where it crept into supposedly scientific explanations of behaviour. In his earliest (almost purely observational) papers, there is as yet no sign of any revolution against the conventional chain-reflex theory of behaviour. Any move away from this would have seemed at the time a concession to vitalistic ways of thinking. Later, his continued rejection of Vitalism was implicit in his search for a mechanism for instinct. A vitalist would either deny the existence of instinct because it conflicted with his idea of free will, or would dignify it with a capital 'I' and require no further explanation beyond the statement of its existence. More recently Lorenz has been inclined to treat Vitalism lightly: 'You can explain anything by assuming an external natural force, but if that is legitimate in science, then the next time my grandson asks me, "What makes the locomotive go?" I shall answer "My dear, this is done by the so-called locomotive power" — period! And if my grandson is content with that answer I shall disinherit him.' At which he roared with laughter.

But in his time Lorenz has taken Vitalism seriously enough to write a philosophical paper on the subject. 'Inductive and teleological psychology' was published in 1942 and is in the first volume of his collected papers. (Teleology is 'the interpretation of biological structures in terms of their purpose or use', and its danger is that it can lead to circular arguments.) This paper is actually his defence against an attack on ethology by a vitalistic student of behaviour, but in it he counter-attacks the idea of animal or human behaviour arising from other than natural causes.

A concept which appears to be Vitalism reincarnate has regained some ground in recent years. This is the idea that the whole of a system is greater than its parts, resulting in a doctrine called 'Holism' (the 'w' being lost on its reversion to Greek as wholeness is elevated to an 'ism'). Lorenz attacked Holism in 1942, and again in 1950, in 'Part and Parcel in Animal and Human Society', an updated and shortened version of a 1943 paper. In this he noted that holists believe in a miraculous 'whole-producing' factor, neither accessible to nor requiring a causal explanation. In their view it is neither necessary nor scientific to attempt to disentangle the interactions between the sub-components. In his 1950 paper, Lorenz was writing about organic entities such as animals or man, but more recently Holism has been invoked again in the context of environment. Certainly, complex systems differ from simple ones in that they have a disproportionately higher number of interactions, thereby creating possibilities that did not previously exist. But this does not absolve the scientist from using his established methods of taking the system apart bit by bit and attempting to analyse it, or of finding some alternative objective method of study.

In objecting to Holism, Lorenz says that it is based on a misunderstanding of another idea that has been of great interest to him, to do with how we perceive order in apparently random events, how we see a pattern emerging from chaos, and how an animal receives, recognises and acts on a complex signal but disregards irrelevant information. In this the whole signal or pattern has a significance that is absent from its individual components. Lorenz points out that it is wrong to generalise from principles about the perception of wholeness and apply them to the nature of wholeness itself.

His European cultural background led naturally enough to his clash with Vitalism, and fitted him well to debate it. But Behaviourism was more remote and, at first, more difficult to take seriously. While he was in England in 1949, in an address to the Society of Experimental Biology, Lorenz dealt with some of the rivals to ethology, dismissing Behaviourism and its proponents almost contemptuously. He subsequently recognised that the paper had a disagreeable 'mother-knows-best' air that did neither his opponents nor himself justice. Two years later in a lecture in New York (included in the second volume of his collected papers), he treated the behaviourists with more respect — in the form of a far more carefully reasoned opposition that nevertheless appeared as an act of provocation that was a precursor to the 'war'. Tinbergen's *Study of Instinct*, giving the detail of their joint theories, was published in the same year.

Two leading American studies of animal behaviour at that time were by T. C. Schneirla and Daniel Lehrman, whose work with and observations on army ants and ring doves, respectively, have been described as being in the same class as Lorenz's work with geese and Tinbergen's with herring gulls. Both were critical of the European ethologists, and when Lehrman produced 'A Critique of Konrad Lorenz's Theory of Instinctive Behaviour' in 1953, the battle was joined. The paper began with a considered reappraisal of the greylag's egg-rolling behaviour, and moved on to cover a wide range of patterns of action which Lorenz and Tinbergen had described as innate, but in which (Lehrman argued) development or maturation might play an important role. He also threw in a few acid comments on the 1940 paper, and ended by listing the 'serious flaws' in Lorenz's theory. The sharp attacks on both Lorenz and Tinbergen clearly exposed the weaker points of the 'instinctive' position at that time, and Tinbergen cautiously and perhaps wisely conceded ground, but thereby left his friend, who was disinclined to retreat so far, exposed to the main

Durch Domestikation vrerursachte Störungen arteigenen Verhaltens

Von

KONRAD LORENZ ´Altenberg)

Mit 35 Abbildungen

I. Einleitung und Aufgabestellung

Die maßlose Verallgemeinerung, die die PAWLOWsche Lehre vom bedingten Reflex durch manche ihrer Anhänger erfahren hat, kann zu Anschauungen führen, deren praktische Anwendung auf das Gesellschaftsleben des Menschen keineswegs unbedenklich ist. . . . (p. 2)

* * *

Für gewöhnlich wird der Vollwertige auch schon von sehr geringen Verfallserscheinungen an einem Menschen des anderen Geschlechtes besonders stark abgestoßen und es siegt auch bei dem oben-erwähnten Konflikt zwischen positiv beantworteten körperlichen Merkmalen und abstoßenden Merkmalen des Verhaltens meist sein Ansprechen auf die letztgenannten. In bestimmten Fällen aber kommt es nicht nur zu einem Verlorengehen dieser Verfalls-erscheinungen vermeidenden Selektivität, sondern geradezu zu einem Umschlagen ins Gegenteil, zu einem Angezogenwerden durch nicht allzu krasse, aber doch deutlich als solche erkenn-bare Verfallserscheinungen. . . . (p. 61)

* * *

Dennoch muß diese Rolle von irgendeiner menschlichen Körperschaft übernommen werden, wenn die Menschheit nicht mangels aus-lesender Faktoren an ihren domestikationsbedingten Verfalls-erscheinungen zugrunde gehen soll. Der rassische Gedanke als Grundlage unserer Staatsform hat schon unendlich viel in dieser Richtung geleistet. . . . (p. 71)

Above: passages from Lorenz's controversial paper, 'Disorders caused by the domestication of species-specific behaviour', written while he was still at Altenberg, and published in German only in the *Zeitschrift für angewandte Psychologie und Charakterkunde* in 1940.

Note 'Geschlechtes' in the middle passage, meaning sex or gender, and also the reference to the racial basis of the German state in 1940. It was the strongly pro-Nazi style of what should have been a non-political scientific attack and perhaps a little hurt at his companion's readiness for compromise. While the middle ground was being defined, Tinbergen could neither produce a revised edition of his book nor allow it to be reissued for another eighteen years. By then, Schneirla was long dead and Lorenz and Lehrman had developed such a respect for each other that it is possible that they might eventually have recognised how much they could in fact agree had not Lehrman died too. As it was, Lorenz can regret the loss of a worthy opponent.

Behaviourism in its most extreme form denies instinct almost completely; in its more modern form it still has little time for it. It is as though each living creature were born with a blank and open mind awaiting the experience of the world around it to mould the character of its response: learning is all. In contrast to Vitalism (but in common with ethology) it provides a mechanistic explanation for behaviour. In Lorenz's view, Behaviourism can only be understood by looking to its origins as a deliberate alternative to Vitalism: 'When the vitalists preached that "Instinct" (with a capital "I") was an external natural factor, there was a tug-of-war in which the mechanists manoeuvred themselves into the opposite corner of the ring.' The mechanists needed a concept as tangible and successful as the atom had proved to be in physics. In strict imitation of the atomism of physics they sought an 'atom' of behaviour; and the 'atom' that offered itself at that moment was the reflex, and with it the conditioning process that had been demonstrated by Pavlov. For a while, it was considered vitalistic, and therefore unscientific, to admit to any other element of behaviour than this: indeed, Lorenz himself had at first accepted it as an explanation.

From that point, the behaviourists and Lorenz diverged along paths that were set for them by their entirely different choices of scientific method, so that in time Lorenz came to reject the reflex as an essential component of animal behaviour, while the behaviourists, by using it as the sole guide to their choice of experiments, reinforced their own belief in it. Reinforcement does work—it worked well

paper that led repeatedly thereafter to strong attacks on Lorenz—such as Lehrman's 'A Critique of Konrad Lorenz's Theory of Instinctive Behaviour' in the *Quarterly Review of Biology*.

Although Lorenz renounced the Nazi taint and condemned Nazism itself, this evidence of his past association has dogged him over the years, coming to a head once more during the period before the award of his Nobel Prize.

enough on the experimenters themselves — but if it is all that the scientists looks for it may be all that he sees. Lorenz comments, 'In investigating the conditioned reflex, Behaviourism had great success and gained great merit. The error of this whole school is not the way they investigate the contingencies of reinforcement, but the way they don't believe in anything else.' True to Lorenz's assertive style, that is something of an exaggeration, for even the most extreme behaviourists do now admit the existence of something like instinct, although they do not regard it as a major component of behaviour — and in man far less than in lower animals. Lorenz, on the other hand, regards instinct as a major part of any animal's make-up at birth, and he includes man in that assessment. The difference is genuinely one of emphasis and interest, so the two views are not mutually exclusive. But in the strength of their emphases they are about as far apart as they could be, short of actually contradicting one another.

The sort of experiment you might see in a behaviourist laboratory in the years after the Second World War was typified by the Skinner-box demonstration in which a pigeon in the box was shown the word 'peck' and promptly pecked it, and when this was replaced by the word 'turn', the bird turned around on the spot. The pigeon had been conditioned to respond appropriately to the different shapes of the two words; and when it performed correctly it received a reward of food. In this type of experiment the situation is deliberately artificial. Lorenz noted a statement by Daniel Lehrman that most American behaviourists took the view that in their sort of behaviour experiments, questions about the processes internal to the subject that give rise to the observed behaviour are unnecessary, misleading, non-scientific and/or irrelevant. In other words, in these experiments instinct was excluded as far as possible because it would confuse the issue as to whether the experiment itself (the conditioning process) was working or not.

As circus tricks, pupils of Skinner have persuaded macaws to pedal bicycles and hop along on a scooter, and parrots to roller-skate for the 'reinforcing' reward of a peanut. Essentially, this is the same as the dolphin show in which these delightful mammals will demonstrate a wide range of natural and acquired skills in exchange for a fish or two from a sympathetic trainer. In the most remarkable demonstration using this technique, a dolphin was rewarded only for devising and performing a trick that it had not previously done: the result was a virtuoso display of inventiveness. If conditioning can even be used to teach a clever animal to innovate, that says a great deal for both the intelligence of the animal and the power of the method that has revealed it.

At the same time, four thousand miles away at Seewiesen, an entirely different range of experiments was being carried out by Lorenz's pupils. Here is an example of one that was also concerned with a learning process. It was devised and carried out by Wolfgang and Margret Schleidt; their research animals were turkeys.

It had been observed that turkeys show alarm when they see a buzzard—a dangerous but rare predator—flying overhead. There are many objects in the sky, so how do they single out a buzzard from a neighbourhood goose or a passing jetliner, neither of which give cause for alarm? The Schleidts asked two questions: to what extent is instinct involved, and how is this combined with learning? Following the method that Lorenz and Tinbergen had used at Altenberg, they made paper cut-outs of the bird in various sizes and passed them along a wire over the pen of turkeys. They found that to produce the reaction of fear the shape of the cut-out was unimportant, but it did have to travel at the right apparent speed, as had been shown before. The turkey interprets a dark patch overhead moving at a rate of ten times its own length per second as a buzzard at height. But it was also found that if no attack came, the effect of the stimulus was gradually reduced. It appeared that, at birth, all objects flying overhead frighten the turkey poults, but that common birds—or rather, the common combinations of speed and size—soon lose their capacity to produce the reaction of fear. Eventually, only the unfamiliar combinations are left, and these include the buzzard. This result was checked by rearing turkeys in a shed where there were no birds flying overhead, and trying the same stimuli on them.

In this example the bird has certain inborn fear reactions, but learns by experience when these are not required. It is thereby equipped to deal not merely with buzzards but with a whole class of dangerous situations which could include attacks by flying predators never previously encountered by any member of the species. Lorenz, on seeing this work, will have recalled that the jackdaws had a different combination of instinct and learning whereby their young were positively introduced to a knowledge of their enemies. In that case the instinctive element is the 'rattle' cry of the jackdaw flock, while a knowledge of the danger it is applied to is passed on culturally from one generation to another.

Experimenters and observers of the two schools react to each other in predictable ways. B. F. Skinner, a recent leading proponent of operant conditioning, clearly thinks of Lorenz as an 'old-time naturalist', that Lorenzian ethology has turned up little of interest, and that many of its findings can be given explanations that tie in well enough with behaviourist theory. One example is imprinting. Plainly, a young duckling is rewarded by remaining close to or moving closer to its mother or mother-substitute. This reward reinforces its subsequent behaviour. A behaviourist experiment demonstrates that instinctive 'following' is not involved. If it is arranged in the laboratory that the only way a duckling can get close to its mother is by walking in the opposite direction, this it can be taught to do. Says Skinner: 'It inherits a capacity to learn to move in a direction which produces a reduction of distance between itself and the mother.'

Similarly, Lorenz can explain conditioning in terms of instinct. A capacity for conditioning, he argues, is developed in many creatures because it can aid survival. Its general purpose is to mete out reward, in the form of pleasure (or 'feeling good') for behaviour that promotes survival, and to dispense punishment in the form of displeasure or discomfort, for acts which endanger survival. An animal which has otherwise little capacity or inclination to think ahead thereby acquires foresight, and so has the chance to pay a present price (in terms of discomfort or unwelcome effort) for something of potential future value. But this innate capacity for conditioning will only work in the artificial world of the laboratory if the innate mechanism is operated properly.

In another example Lorenz quotes the remarkable capacity of rats for discovering what is good or bad to eat, as has been shown by an experiment where essential constituents of a rat's diet were broken down into their component amino acids which were then presented to the rats as an array of separate foods. The rats quickly learned to eat the right amounts of each to achieve a balanced diet. That they found the combination that made them 'feel good' was suggested by a second experiment in which they were allowed to drink saccharine solution before being injected with a drug which gave them a feeling of nausea, with the result that they developed an aversion to the last thing that entered their stomachs—the saccharine. They could be readily conditioned by nausea to avoid food, but (and here is Lorenz's point) no other discomfort inflicted on them made the slightest difference to the rats' choice of food; they already had an innate mechanism for telling them what is good or bad to eat.

Lorenz calls an animal's capacity for being conditioned the 'innate schoolmarm'. This lady can school her pupil by a combination of punishment and reward: fear of punishment propels him away (in any direction) from undesirable actions, while the promise of reward draws him, from any side, towards a desired response. She finds that an economic balance between the two is the best way to guide him along the shortest path between them. Where this capacity is required for survival it may appear by evolution. The animal may develop a set of special predetermined patterns of behaviour that must always follow particular stimuli (instinctive behaviour following the operation of 'releasers') or an equally predetermined mechanism for producing the sort of response that is associated with subsequent reward (a capacity for conditioning). When the behaviourist method works, which is a great deal of the time, it can work very well, but by the same token it has clear limitations. Since behaviourists cut blindly across species differences, the only laws they find are those where behavioural reactions are common to the different species they have chosen to work with. 'The technique and the philosophy which it expresses carefully avoid a good deal of what makes a guinea pig a guinea pig and a pigeon a pigeon — to say nothing of what makes a person a person': here Daniel Lehrman is appreciatively quoted by Lorenz.

In Germany, Paul Leyhausen (a pupil of Lorenz) looked at the claims of the behaviourists and devised experiments to show that animals really do possess an internal clockwork that is to some degree independent of the external situation, demonstrating that an animal is not in fact a slot machine. And fully-fledged American Behaviourism was not only attacked from without but also mined from within. Keller and Marian Breland were one husband and wife team who began pulling bricks out from beneath the edifice. They were early pupils of Skinner who took his methods away from the laboratory into the 'real world' where success or failure is measured not in the pages of learned journals but in the columns of a bank book — and by this form of accounting, Behaviourism soon began to get into trouble. At first the Brelands worked on a defence project, training pigeons to search out and indicate people who might be snipers waiting in ambush. The success of the project was prejudiced by the fact that in real life the pigeons, however well conditioned to perform their assigned task, would insist on going to sit in trees. The Brelands then decided to train animals (still by means of Skinner's 'operant conditioning' methods) to perform circus tricks, but found that in a significant number of acts the animals were difficult to train, or their conditioning broke down after a promising start. The reason was that the animals would always insist on doing their own thing. For example, their pig, instead of getting up and playing the piano as should any well-conditioned creature, would root around and chew corners off the pig-sized piano—and would do this in the total absence of any reward. Finally, when further reinforcement of the pig proved unsuccessful, the Brelands gave in and, instead, reinforced the piano. 'We've had great difficulty designing a pig-proof piano', said Marian Breland. Rabbits, they found, would nibble at coins instead of transferring them properly and promptly from purse to money box to claim their reward—even though the time spent nibbling slowed its arrival. But it was with the invention of 'Sammy the Dancing Chicken' that they finally gave in. Chickens like to scratch, so a 'dance' makes use of something the chickens would want to do anyway; the fact that it is done on a polished floor while music plays is simply a redirection of the innate behaviour pattern by conditioning methods.

Up to the time of his Nobel Prize (and perhaps even since), where Lorenz's name was known among American biologists it was connected in a rather general way with his plea for recognition of innate, 'environment-resistant' features of behaviour; he was seen as a defendant of the concept of instinct and a proponent of a more naturalistic study of behaviour. But until a selection of his papers appeared in translation in recent years, an understanding of what he actually said on these subjects was limited by a lack of knowledge of his original writings, apart from the popular books.

At the end of the nature-nurture argument, what do we have? It appears in all this that Lorenz has replaced the behaviourists' one 'atom' of behaviour (the conditioned reflex) by two ('instinctive' and 'learned' behaviour). The word 'learning' itself has often been widely and loosely used for several different processes in which information flows from the environment and produces a change in the behaviour of an individual. Lorenz himself often uses the word in a very broad sense to include almost everything affecting behaviour that is not part of the genetic endowment of the individual. But, looking closer, it might seem that any behaviour pattern, even a simple one, is composed of a number of components, some of which are genetically determined and some of which have influenced either from outside or by the animal's own changing physiological state. The behaviour itself lies in a graduated range—from that which is fully instinctive (pre-programmed and environment-resistant) to that where learning has played a large part. Lorenz then simplified this by dividing behaviour into 'closed' and 'open' programmes. In a closed programme, the genetically-determined pattern emerges virtually intact, though usually on a given signal from outside. (It will also be affected by physical factors such as heat or cold, or whether the chemical constituents of the animal's environment lie within a certain range; but this does not affect the general argument.) Open programmes are those which depend for their development upon additional information coming in.

To an English mind (and indeed to some of Lorenz's own colleagues and pupils), this seems a rather rigid way of looking at things. English-speaking ethologists have tended to see behaviour as a continuum in which environment-resistant behaviour patterns are at one end, and those depending strongly on learning are at the other. It is recognised that the survival of a species will often select for and then depend upon genetic programmes that are highly structured, so that an animal will be in no doubt as to what to do in a whole range of situations where its survival is even marginally in question. In other cases, such precise information would be unable to achieve the result required (as with Schleidt's turkeys and their flying predators), so a mixture of instinct and learning is selected. Another type of programme appears to require the individual to apply trial and error to the environmental situation and do that which rewards it by achieving the programmed objective. Overall, this creates a situation very much like the 'parliament of instincts' proposed by Lorenz; many of the instructions will conflict and action will depend on how the balance tips in favour of one or another, or whether some compromise is possible. It is easy to imagine closely balanced but opposing instructions building up to a high intensity of indecision, or even becoming locked in some extreme condition, as actually happens in both animals and man. To most 'English-speaking ethologists', this graduated range fits the observations of Lorenz (and for that matter, the behaviourists) quite as well as any rigid division into clearly defined categories. And, in practice, many of Lorenz's own colleagues and pupils seem to accept it too. Because it does not demand or seek firm answers to what may be an unnecessary question, they would regard it as a better starting point for the study of behaviour. The question 'What parts are innate and what learned?' can be either by-passed or studied for its own sake.

Jürgen Nicolai, who joined Lorenz at Buldern and was with him throughout his time at Seewiesen, has studied the contributions of inborn and learned characteristics to the song of birds. A fascinating, complicated, yet revealing system to examine is the relationship between birds that, cuckoo-like, deposit their eggs in the nests of other host species. Nicolai has studied African widow birds of which the young can mimic the host nestling (a finch) almost perfectly: their cries and begging movements are the same, and the deception is complete even to the markings on the inside of the hatchling's mouth. But there are, in fact, some seventeen species of widow bird and each has to lay its egg in the nest of the right species of finch: this is ensured by the behaviour of the male widow birds which learn the song of the appropriate host while young, and include this 'foreign language' as part of their own repertoire.

Part of Nicolai's experiment was to take an 'artificial' host species (the domestic Bengalese finch) to Africa, where it would not normally be found. He then collected eggs of widow birds from the nests of their chosen African hosts and reared them under his artificial host. As the widow birds grew up they developed a song which contained the normal inborn elements of widow bird song (which they had never heard) together with elements from that of the host species (which their own species had never heard before). Here inborn and learned existed together as part of a single, complex, repeated pattern of behaviour—the composite song occupying a middle position in the continuum of behaviour.

Lorenz accepts this as an excellent piece of work, in part because he knows that Nicolai has studied the whole animal and not just this one aspect of its behaviour. So what does he mean when he says 'there is no third way'? It becomes clear that Lorenz is distinguishing not between 'innate' and 'learned' behaviour (which are so difficult to define), but simply between internal and external sources of information: the messages that are encoded into a creature's genetic system, and those which come from its environment. Lorenz does say this himself, but it must be admitted that it is not always clear from his writings. It seems that the way a researcher classifies behaviour depends strongly upon his own interests. Lorenz is primarily interested in how an organism can fit its environment, like a lock and key: he is interested in how this fit is achieved, what parts are programmed into the species and what are due to adjustment by the individual. Beyond that, he is interested in how it came about, and in the process of evolution itself.

To some English ethologists, and to Robert Hinde in particular (the Director of the MRC Unit on the Development and Integration of Behaviour at Cambridge), this way of looking at things is less helpful. Hinde's Unit was founded upon his interest in how behaviour develops in the growing individual. What, he asks, makes one individual different from another; and can we discover the source of these differences? These questions can be investigated by taking genetically identical pairs (or, where this is difficult to achieve, genetically similar groups) and placing them in environments which differ from each other, or, alternatively, by taking genetically different pairs and subjecting them to identical rearing conditions. Hinde describes his method as depending upon 'the dichotomy of the source of differences', as distinct from Lorenz's dichotomy of the sources of information; or the dichotomy of behaviour itself which, as we have seen, exists only in a superficial sense. He also identifies a fourth type of dichotomy: sometimes it is convenient to distinguish between on the one hand the consequences of 'maturation' (the tissue growth and differentiation that causes an organ to develop to perform its function), and on the other the consequences of experience (the signals that reach it from the outside world). Hinde emphasises that students of behaviour should be clear about which type of dichotomy is being used in any case, and should not apply a dichotomy which is suitable in one situation to another where it is inappropriate; and should not mix the various categories in a single comparison. (If the reader finds this difficult to follow, pity the poor biologist who must live among all these landmines!)

Such differences in approach do not matter too much if, in the end, the same research gets done (this seems to be Tinbergen's position). But there are more specific points of conflict; even, for example, over how the word 'innate' should be defined. In addition, Lorenz is concerned that in what he regards as an ill-conceived attempt by English-speaking ethologists to reconcile ethology with Behaviourism, his concept of spontaneity seems to have been forgotten. His 'hydraulic model' in which an unsatisfied drive builds up and seeks an outlet remains powerfully in his thinking. There are a number of matters of detail in which Lorenz and Tinbergen now take differing views; Hinde's influence is also very strong in England.

But specific issues apart, the main difference between Lorenz and the English group is one of interest. The Germans have a splendidly sonorous word for it: 'weltanschauung', the philosophy of life that you derive from the picture of the world as you see it. If, as many people seem to do, you apply such a philosophy back to the world you act upon, you complete a circle which is self-reinforcing: an 'ism' is created. Certainly, Hinde and Lorenz have different world pictures. A way of illustrating this objectively has been applied by Wolfgang Schleidt: he compared the references in textbooks by Hinde and Eibl-Eibesfeldt (who follows Lorenz closely). There is some overlap, but the differences are significant. Hinde reads (and therefore quotes) more papers in psychological journals, while Eibel covers the German literature better: but there are other differences simply in the journals that they read and can therefore quote, even when working on similar topics.

Let Lorenz have the last word in the debate within the discipline that he did so much to stimulate. It is one which goes back to the start of this discussion, and extends all the way through it:

> I am rather bitter about philosophy. Philosophy has done a lot of damage by biasing people, making them quarrel about matters which cannot be settled. If you see men of genius who are obviously blindfolded with regard to certain questions, it would be very supercilious of us to assume that we ourselves are not blindfolded with regard to questions which will be obvious to everyone tomorrow.

If we now turn again to human behaviour, we find that Lorenz and many of the English-speaking ethologists (Skinner, too, for that matter) can be grouped together as having one major characteristic in common—their emphasis on the animal inheritance of man which arises from having based their earlier research on animals. In addition, when they do begin to study humans, they think first of the methods that worked with their animals. But distinct from both Lorenz and Skinner were the scientists who based their research on the special qualities of man. Looking back, we can see that these too can be divided into opposing groups. Ranged alongside the Skinnerian behaviourists were assorted Marxists, sociologists and humanists (such as Ashley Montagu) who believed that man was born with very little pre-structured behaviour, and that environment and learning were by far the most important part, if not all of the making of man. Parallel to Lorenz were those who saw special inborn human qualities. The French 'structuralism' of Claude Lévi-Strauss in the 1950s was one expression of this, and another was provided in America ten years later by Noam Chomsky, who proposed pre-existing structures for the grammar of human language.

In England the seeds of compromise were sown long before — by J. B. S. Haldane, who as a geneticist could see no way of separating nature and nurture as the opposing factions would wish. In that tradition, many of the psychologists and physiologists investigating child development have concentrated on the interaction between heredity and environment. And alongside that came spectacular advances in the investigation of brain physiology and its role in mediating human behaviour. In America, these other sciences developed separately from orthodox Behaviourism, and by-passed its more rigid forms to set the scene for the 1970s, when scientists studying human behaviour could agree to approach it from many different angles, with each method (now quite properly including those of both operant conditioning and ethology) making its own contribution.

But before that was to come about, there was one big battle still to be fought — that to do with aggression itself.

Chapter 11

Aggression

Throughout his work Lorenz emphasises the role of instinct, and the instinct with which he is most closely linked in the popular mind is aggression, particularly its expression in man. It is seen as a drive in all of us that seeks an outlet in one form or another. It can be directed into harmless channels, but if simply suppressed, the argument goes, then sooner or later it will burst forth with explosive force.

In this view of Lorenz we already have one misconception, a misunderstanding of what he was trying to say for which Lorenz himself must bear some of the responsibility. It arises from the English title of his book, *On Aggression*. The German original was *Das sogenannte Böse*, literally 'the so-called evil'. In German this is a good title: 'sogenannt' being a much stronger word than its English equivalent. Perhaps because the German language is more adapted to the precise expression of meaning, the evolution of a word or its use in an unfamiliar or inappropriate context calls for 'sogenannt' — to imply that the qualified word should be seen as placed in quotation marks. Lorenz, not surprisingly, is very fond of 'sogenannt'; in a lecture he may emphasise his use of analogy still further by drawing quotation marks in the air before and after speaking the word he wishes to qualify.

As we have seen, he values the common use and meaning of words as an indicator of the unity of the structure they describe. For example, if in man we see something we can call 'jealousy', we are describing jealousy as a unit of behaviour which is as real as a hand or an eye, a unit of structure. If Lorenz then sees a similar form of behaviour in an animal, he feels justified in calling this by the same name, 'jealousy'. This does not mean that the animal and the man have a common ancestor from which the jealousy behaviour was inherited: it is a purely functional definition. We accept readily that in the physical organs of the body, analogy of form means sameness of function.

As Lorenz puts it: if he dissects an octopus and discovers an organ with a lens, a diaphragm and a retina, in short a camera eye, he can call it an eye at once without ever working with a live octopus to see it in action. He does not have to excuse himself, saying that he realises that cephalopods developed eyes independently from vertebrates. In this case it would obviously be ridiculous to call one 'the analogy' of the other, for both are the real thing, both are true 'eyes'. Similarly, the jealousy that he sees in geese is not just an analogy of the jealousy behaviour of man but an expression of some common underlying structure. Goose and man have quite independently evolved the same structure, the same 'organ', and again, both are true examples of the same thing. To Lorenz, this is a tremendously important argument. In his Nobel Prize lecture he emphasised that 'no such thing as a false analogy exists: an analogy can be more or less detailed and hence more or less informative'. As a result, he feels justified in using a common term for the aggressive acts that he sees performed by many animals and also by man. Here, too, there can be no 'false analogy' between the two if they are functionally the same.

One problem with this is simply that it is much harder to get observers to agree that a particular pattern of behaviour exemplifies 'jealousy' or 'aggression' than for them to agree that a particular structure may be called 'an eye'. In the view of some ethologists, notably Robert Hinde at Cambridge, Lorenz's argument substitutes the naming of things for the identification of soluble problems. But Lorenz himself is satisfied that he has identified a real instinct, and one that should be the urgent objective of long study until such time that, individually and socially, man can control the evil of misdirected human aggression.

But as the German title implies and the English title does not, 'the so-called evil' instinct for aggression is not necessarily bad; unless it is misdirected, it serves a positive purpose. Lorenz emphasises that aggression is one of many instincts that are in competition in the healthy animal that is living a normal life. This 'parliament of instincts' is an organic structure, the balance of which may be tipped one way or another depending upon the state of the animal and its interplay with the environment. In man, too, there exists at a deeper level than in the animal this same parliament of instincts in which aggressivity should play its normal, healthy role. The problem, he tells us, is that in the cramped and crowded conditions of modern life, natural outlets are suppressed and the drive builds up dangerously.

On Aggression is a bad title for the book, and not just because it omits the implication of a positive purpose. Quite simply, 'aggression' as most people understand the word is not what the book is about. In its common meaning, aggression equals violence, which is not the instinct itself but may be the end product of its suppression. Years after, this English title nags Lorenz like an aching tooth: 'It ought to be *On Aggressivity*', he told me emphatically. But this, of course, is a weak title in English; it promises an erudite treatise remote from the broad public interest that the book certainly has. 'Aggressivity' would have halved the appeal of the book in two extra syllables, while the more 'aggressive' salesmanship of the crisp but imprecise title allowed Lorenz to be heard and still sells more copies than the timeless, gentle *King Solomon's Ring*. It was so successful, in fact, that it led to a spate of popular books on the animal in man. Robert Ardrey, already roaming in the fields of man's *African Genesis*, followed it up by an investigation into his *Territorial Imperative*. Desmond Morris, formerly curator of mammals at London Zoo, was so successful with *The Naked Ape* that he resorted to voluntary exile in the temporary tax haven of Malta. Two other zoologists with the unlikely (but genuine) names of Robin Fox and Lionel Tiger also placed their joint claim on the territory with yet another book on man, *The Imperial Animal*. If Lorenz felt alarmed by some of this, was he partly motivated by an instinctive reaction for territorial defence? After all, this was originally his own private field, and here was a host of intruders each offering the public his own interpretation of what ethology said about man.

As it turned out, the interest of the public was great enough to absorb all of these and more besides; but Lorenz still felt concern at the way ethological ideas were treated. In the introduction to his collected papers he writes, 'Some of my allies make me squirm'. Desmond Morris, 'who is an excellent ethologist and knows better', made him unhappy by emphasising the beastliness of man at the expense of his capacity for conceptual thought. Robert Ardrey's scientific popularisation excites his admiration—until it is Lorenz's ideas that are being popularised, when he begins to perspire like the passenger in a car that is being driven far too fast. Since then, he has made his peace with Ardrey, whose later books he finds more level-headed, while Tinbergen more than recognised the virtue of Morris's work by welcoming him to a place in his Department at Oxford. And in America the imperial Tiger has been called in to advise the Harry Frank Guggenheim Foundation on the distribution of funds for research into the roots of violence and related matters. (Lorenz approved of

Fox and Tiger from the first: when their American publisher sent him a copy of *The Imperial Animal*, he delighted the authors with a letter of praise that began 'My dear Carnivores,' and also contributed an introduction.)

The book that opened the floodgates was originally conceived as a more substantial and learned work for which *On Aggressivity* would, indeed, have been the appropriate heading, but when the book's broader interest was realised, Lorenz came under pressure to sharpen his statement. He went as far as he could with this and, as it stands, there are still eleven chapters on animals before the reader is invited to follow his extrapolation to man in the final three—and of these the first is 'On the Virtue of Humility'. He has said that even this does not state as emphatically as he would wish that 'until this point everything I have said is hard science; from now on I speculate'. In fact he does say something very like that at the start of his chapter twelve, but the crisp and assertive way in which he then makes his speculations on man is the signature of a writer who, while lacking scientific proof, has no real doubt that he is right.

When the book was published it was the extrapolation to man that struck a sympathetic chord in the public consciousness—as his publishers expected. Lorenz has been under attack for his real or supposed statements on human aggression ever since; and the antagonism of anthropologists and psychiatrists (particularly in America) was aroused by 'his often wildly speculative forays into the realms of human culture and behaviour' (to quote Daniel Ehrenfeld of Barnard College, New York). His newly publicised ideas added fuel to the controversy that was already growing in the more narrowly scientific field.

Why is he attacked? Lorenz himself believes that it is because of the spiritual pride of man who is prepared to accept that his intestinal tract is a product of evolution, but resents the suggestion that his standards of social behaviour are also in some way the product of an evolution that can be studied by biological methods. As for aggressivity, he says this has got itself a bad name by its identification with authoritarian military aggression. Man prefers to regard himself as rational, with few or only weak instincts; and of all those that might be attributed to him, the most unpopular is aggressivity. Lorenz feels that his view is not accepted because of an ideological limitation of man himself. But the very success of *On Aggression* and all the other related, popular books points to an exactly opposite picture—that of man wallowing in the realisation of his own animal heritage, or accepting it as an explanation for the actions of those whose behaviour cannot otherwise be explained. The attacks came not from the broad general readership of these books, but from some of those with a direct intellectual interest in the subject. Lorenz has a fall-back position to deal with this: he says that what his critics really object to is that biology has overstepped the limitations that they would set for it. In other words, this is not a purely intellectual discussion: it is a contest for ideological territory.

In his animal work Lorenz is on home ground: in his book he tells many stories detailing aspects of animal life in which aggression plays a part and in which it is related to other instinctive forms of behaviour. But what was missing from his first draft was a simple, concise statement of what aggression is and what it is not. Instead of making generalised statements about aggression he proceeds by example, detailing many of the observations that he and his fellow ethologists have made. He offers no preconceptions—this, he claims, is how he himself works—but expects the reader to draw his own conclusion, as Lorenz himself does at the end of the total experience. It is an interesting way of presenting his material, but it differs from real life. Firstly, nobody really starts without preconceptions. We all measure each observation that we make against previous experience in some degree; otherwise we would be totally unable to separate what is significant in the scene before us from the random, meaningless or irrelevant events. In the book, Lorenz offers us his own selection of observations together with his partial analyses, but even so it would hardly be surprising if the reader made a different synthesis. As we have seen, many of those with previous experience do, while those without have little choice but to accept the conclusions offered.

The orthodox and more readily understandable way of writing would be no more artificial than that chosen by Lorenz. It would first lay out the basic idea or hypothesis (which is generally the conclusion to which the writer has already come) and then it would present the evidence: that which supports the conclusion, that which is consistent with it, and, unless the writer is to risk the charge of special pleading, that which requires further explanation. At the end the reader should be able to judge for himself how valid—and how selective—the evidence is. Lorenz rejects this approach because he wants the reader to feel his own sense of excitement in learning by discovery. At this level, as in *King Solomon's Ring*, he is remarkably successful.

In a poetic opening chapter he floats weightless over the coral reefs south of Florida, imagining himself as a naive observer of the various species of fish. Some pass him in vast, anonymous clouds; others, generally more brightly coloured, swim as individuals or in pairs. From the dazzling disorder of the coral world, a pattern is gradually abstracted by his mind from the balletic play of darting forms, and erupts as an insight into the role of the brilliant poster-paint colours as warnings to other fish with the identical bright display to keep clear: on its own territory, the bearer will back its threat with a will to fight that would almost certainly drive off any intruder. Lorenz then discloses that he was, in fact, by no means a naive observer; he had gone to America partly to confirm a hypothesis gained from several years of working with tropical fish in tanks (dating back, in fact, to Seitz's work with cichlids, and his own even longer observation of other creatures). The fish in the wild did indeed behave as he had expected from his earlier experience.

The chapter also reveals a shift of emphasis in his animal interests. Having set up his post-war colonies of geese and ducks, Lorenz was gradually allowing these to be taken over by his pupils, while he himself had once again been captured by the quiet beauty of the fish tank to which the extravagant tropical diversity of forms and colours added a new dimension. From his observation, so many years before, of the long-drawn-out ritualised fights of the cichlids, he was attracted to the study of their aggressive instinct, and from this to aggressivity in general, asking himself, as always, 'what is it *for*?' He was now equipped to make comparisons of aggressive behaviour in many species which he or his friends had studied. The range of animals covered was wide enough to suggest that in at least some of them it must have been developed independently by separate evolutionary pathways. Where evolution fashions an organ or element of behaviour time and again in different species, this convergence means that it is potentially an important aid to the survival of species, even though all creatures may not have it. Organs for seeing or for grasping are convergent in this sense; so are organs for flying, though in a smaller number of species; and so too is this organ of behaviour, aggression.

Lorenz devotes a further chapter to examples of aggression which serve a positive purpose, going on to discuss its role in the preservation of species. He gives examples to show that it is innate and grows spontaneously until discharged. (This is controversial, as we shall see. Other ethologists offer alternative mechanisms which produce much the same results.) Lorenz continues by discussing the factors which control aggression, its expression in ritualised forms, and its channelling along harmless paths. At this point in the book he broadens his approach to write more generally on the way in which a variety of instincts meet within the individual, the balance tipping in one direction or another to prompt the action of the moment. He shows also how this leads to forms of social behaviour which appear remarkably like moral or ethical systems.

Growing out of this is a very important part of his story—a description of the various types of societies that different species form. One is the anonymous crowd: where no individual recognises another there is neither individual friendship nor enmity. In this society, aggression within the species has no function and so is not provided by evolution. A second society is the colony which defends the territory that is common to the group: his example is herons. A third is the tribal grouping, in which there may be no individual recognition of others, only a knowledge of friend or foe, identified by familiar or unfamiliar tribal markings. The example he gives here is the 'large family' of rats, which have exemplary social behaviour towards other rats with a familiar smell but unite in bitter antagonism towards the rats of other tribes. In the fourth type of society there are bonds of friendship or love between individuals. His main example here is the greylag goose—and this chapter is long, because he has much experience on which to draw. It was his closest approach so far to writing a goose book for a popular audience: geese here play the role that was filled by jackdaws in *King Solomon's Ring*.

The general rule that emerges is that aggression may exist without love or friendship, but the bond between individuals does not exist in societies without aggression. There is a further point that should also be noted: the four types of society described are extremes; a single species may have more than one of these characteristics. Lorenz found an unusual combination in crescent-shaped coral fish called 'Moorish idols'. Individually, they would defend their own territory in a corner of the tank, while in free water they would school, exhibiting a rank (or 'pecking') order. Goose society, too, is a mixture. It is based on pair-bonding (love and marriage) combined with group territorial defence—and so is man's, although not necessarily in the same degree or form.

The misunderstanding of *On Aggression* was due in part to the lack of a common language between the various groups of scientists that study behaviour, to its own title and form, and perhaps even to what it does not say concisely and explicitly. One of the most important single statements about aggression as Lorenz understands it is a largely negative one: while the attack of one species on another may exemplify aggression, in practice it rarely does. The predator feels no animosity for its prey; it simply wants food. And the kitten that plays with a half-dead mouse is exploring its own prey-catching movements. Nor is there aggression in the trapped victim of an attack fighting fiercely in its own defence (the 'cornered rat' response).

A display of courage for the benefit of a partner does presumably contain a measure of aggression. Lorenz was delighted by a film he had of Niko Tinbergen being attacked by an Andean goose. The female had incited this by a feigned attack. The male then took his partner's cue and charged in furious anger on the intruder, striking him with strong and painful blows from the wings and bites with its beak, until the vanquished Tinbergen fled the scene, 'black and blue'. This was followed by a vigorous triumph ceremony between goose and gander wherein they expressed their mutual admiration and pride, thereby reinforcing the bond between them.

Was there a tinge of aggression also in Lorenz's laughter at his friend's discomfiture? If so, it could again have been a factor in my own delight in filming Lorenz himself being attacked by one of the snow geese that some years earlier had been imprinted on him. Lorenz wished to examine the nest for unfertilised eggs, but he too was battered and bitten. Yet the behaviour of the animal itself in defending its nest against a predator of another species was not true aggression: a second instinct had overcome the effects of imprinting which should have allowed Lorenz, the 'mother', an unmolested approach. The forebears of the snow goose had known no predators larger than the arctic fox, which they could drive off successfully — so by its ancestors' bequest of courage it could turn on Lorenz too. But if his greylags, programmed by their heritage of lost encounters with red foxes and other larger and more determined predators, had been put to the same test, they would have resorted to a course of noisy distraction, seeking to divert him from his supposed attack by drawing him away from the nest.

If the borderlines of aggression are hazy, some of its positive purposes are clear. They apply mainly to interactions between animals of the same species; that is, they are intra-specific. Their survival function is to space out individuals (or pairs or groups) over their available habitats to ensure the most favourable exploitation of a region and the food sources it contains. Different species have different needs, which means that a whole range of animals and plants may occupy the same territory without infringing one upon the other. There should be no attacks between species other than those between predator and prey, or perhaps in a dispute over prey that has already been taken as food; or some other resource such as a nesting site that is equally valued by two species: although such competition rarely leads to a real fight. The balance between species can be dramatically altered when a new animal arrives on the scene—but still without fighting. A more efficient predator may dispose of its rivals by competing successfully for the same food: Lorenz quotes the spread of the dingo (wild dog) in Australia as an example.

But even within an aggressive species there is little actual fighting, for the mechanisms of aggression are adapted to the resolution of conflict, not to the maintenance of standing feuds. Once an animal has successfully laid claim to some territory, its ownership is recognised. Lorenz described the activity of the poster-coloured coral fish whose success in a territorial contest is closely related to the distance from the boundary of their territory. The intruder will generally give in very quickly, but if the victor carries the chase to his neighbour's home ground, the loser will regain courage and the former victor be put to flight. The territorial boundary is rapidly and clearly defined, and remains invisibly fixed in position until one fish or the other becomes so weak or lazy as to give up part of his territory.

But what happens if the ordained loser is prevented from flight? I saw this demonstrated beautifully at Seewiesen when a caged shama bird was brought to the window of Lorenz's study from which his own pair flew free to the limits of their territory in the surrounding woods. The 'owner' of the territory swept down from the trees and back to the study in an attempt to attack the caged intruder, while his mate encouraged him with her intimidating territorial song. The male flew repeatedly at the wires barring his path, fluttering and struggling to get at his opponent. To the intruder the cage meant not safety but an agony of confinement, the inability to flee from an unwinnable battle. So the contest continued until, finally, in compassion, the cage was carried away, its occupant a panting and pitiable figure by comparison with the proud victor who had expended just as much energy but now trilled his ascendancy with no sign of fatigue: the presence of a wall separating the fighters had only made the contest worse. And in that we have an analogy for modern man. With his machines of war a human soldier, sailor or airman may no longer be fighting hand to hand, but can strike at a distance against an opponent who has no means of showing his distress or desire to withdraw from the conflict.

Lorenz notes the case of rats which may engage in warlike killings within their own species. Ants, too, may go to war, as may hyenas; while lions and certain monkeys may murder their own kind: when a new, dominant male takes over the pack he slaughters the offspring of the previous leader. Most of these observations have been reported since *On Aggression* was written. That their discovery should have been so delayed only serves to illustrate Lorenz's point that when aggressive animal opponents meet, little or no fighting is normally necessary: a mere show of readiness for battle is generally enough. Between coral fish, bright colours are part of their threat. Territorial birds may have this, too, and they emphasise it by song. Poets (including Oscar Wilde) have written rhapsodies over the song of the nightingale, supposing its beauty to be feminine. Lorenz scorns a poetic imagination that flies so ludicrously wide of the mark, for the song is in truth the battle cry of a tough, aggressive male.

A second use for aggression is to determine social order. When I threw an easy catch of mealworms down for Otto Koenig's free colony of cattle egrets, their leader made himself known as soon as the supply began to diminish. By ruffling up the feathers on the back of his head and turning this way and that he quickly persuaded his followers to move back to a safe distance, and well away from the remaining worms.

In many species the complex social hierarchies are set up in early life. An individual's dominance is highlighted not by a show of active aggression but by its near absence: again, a minimal display achieves much. Often it is the middle-ranking animals that offer the most active aggression. They have more to gain than those who are above them, and more to lose than their inferiors. By right, the leaders have the first choice of food, maintaining their strength for defence and providing the best stock for breeding: again this favours the preservation of the group. Prior to mating, the young adults may parade their aggressivity; the young greylag gander will go out deliberately looking for trouble. The 'triumph ceremony' of Lorenz's geese consists of aggressive gestures directed past (or sometimes alongside) the partner; among his Canada geese it contains explicit elements of aggression, shown when they bite each other a little. Except for the relatively rare species which condone lethal wars or murder, when aggression does reach the stage of fighting, the contest is almost invariably conducted according to strict rules designed to produce a clear winner with minimal injury to the loser. The cichlid fish which helped to build Lorenz's interest in aggression demonstrate this well.

As the young cichlids grow up together, they have a pattern of stripes. One fish will eventually assert his dominance by changing this to a line of spots, together with display of a diagonal red bar across the eye. In a small tank, only one male at a time can so claim authority without conflict, but if he is removed another will almost immediately change his stripes to spots. If two mature fish choose to dispute a territory they will circle each other so that the spots and size of each can be seen broadside on by the other. If neither has conceded at the end of this display, they meet mouth to mouth to grasp each other and pull. Their intention is not to cause damage by biting, although a long contest may leave their mouths tender for a time: they are simply trying to exhaust each other. After a while they will release, circle and display once again; then, with neither still conceding, the tug of war will be resumed. This vigorous activity can continue for as much as an hour when both fish are strong, until eventually, during a display episode, one will admit defeat by changing his spots back to stripes. The pattern is in the flesh beneath the scales and it changes with dramatic speed, literally in a matter of seconds. The loser has taken on the appearance of an immature cichlid, and the winner is no longer able to attack him. This change is called an appeasement signal, and it results in the immediate inhibition of the other's aggression. In cichlids the fighting is no mortal combat; rather, it has become a ritual.

Such aggression-inhibition signals are deeply rooted instincts: they have to be. The domesticated muscovy duck has a degraded social behaviour in many respects, but once when I filmed two drakes fighting for the favours of a female, the submission signal was clearly seen on a single frame at the end of the fight. For barely a twentieth of a second, the loser extended his neck low along the ground, and then he ran. The winner devoted another second or so to driving him away, but immediately returned to claim his prize, the excited female who accepted him at once. This followed the episode in which the same animal courted Schutz's boot; it was this courtship display which had aroused the female, and the fight had started when the drake turned to investigate this alternative attraction and found another male disputing his interest. There was clearly nothing wrong with these birds' more basic drives nor with the sign language that accompanied them.

Aggression inhibition can operate in curious ways. When they were working together at Seewiesen, Wolfgang and Margret Schleidt found that it plays an important part in the brood care of turkeys. They wanted to find out how a turkey hen can recognise her newly-hatched young and distinguish them from other living or dead objects around the nest. From an earlier experiment they had some poults (young turkeys) that had been deafened surgically on hatching. When these had grown up, the Schleidts found that the deaf hens mated and laid eggs quite normally: hearing was not crucial to their behaviour so far. The hens sat on their eggs until they hatched — but then, quite unexpectedly, all the deaf mothers pecked their own hatchling poults to death.

Turkeys attack everything unfamiliar (it might be a predator) but it appears that they have no innate image of what their young should look like: they are normally prevented from attacking them by the poult's characteristic cry. That this crying was an aggression-inhibition mechanism was proved by a further experiment with a stuffed stoat, which any turkey mother would normally attack on sight. Inside the stoat skin the Schleidts fitted a tiny loudspeaker which reproduced the poult's cry. Now it was the turn of the turkey hens with normal hearing to behave abnormally: they were no longer able to attack the potential predator.

The most important aggression-inhibition mechanism is that provided by the bond between territorial individuals. 'Love and marriage' overcome the aggressive instincts of both male and mate. The bond of loyalty or friendship between comrades can also be strong, but among his greylag geese Lorenz has seen that when friends quarrel, the conflict is a great deal more bitter and prolonged than a fight between strangers. It is also more likely to result in actual injury, for example by the loss of primaries (the main wing feathers used in flying). This corresponds to human experience that murders are often committed by 'friends' or marriage partners.

According to Lorenz, aggression arises spontaneously and seeks a discharge; it is a matter of observation that this is so, and it must therefore be purely instinctive. Robert Hinde disagrees emphatically: Lorenz's statements, he says, 'do not bear examination'. There are many events that occur during the development of the individual which influence its eventual behaviour. Lorenz, he says, offers no adequate evidence for the 'extreme hypothesis' that aggression will continue to grow and burst forth in the absence of external factors which normally precipitate it. And if there is no good evidence for Lorenz's hypothesis, then 'opinion is against it'.

Attempting to arbitrate between these two apparently incompatible views, Tinbergen comments that both of his fellow ethologists have made loose statements and so have misunderstood each other when in reality they are close to a common opinion. Lorenz and Hinde are both aware that fighting is started by a mixture of internal and external variables, that it is not just a penny-in-the-slot reflex response. Hinde says Lorenz is thinking of a fighting reaction as a short-term response in the developed animal: clearly there must be an intervening process which transforms the undeveloped individual with the innate information into an animal capable of active aggression. Hinde, on the other hand, is very much concerned with the process of development itself, so that he can point to factors which are 'internal' at the time of the aggression but were 'external' at an earlier stage of development. What it boils down to is that Lorenz and Hinde simply have different points of view, and unfortunately from their different stands, neither Lorenz nor Hinde could accept Tinbergen's adjudication; each remains satisfied that his own position on the development of aggression in animals is the correct one.

It was Lorenz's last three chapters—on aggression in man—that loosed the main storm about their author's head, though in retrospect they seem mild enough. He starts with his sermon on humility; he demotes man from his self-proclaimed position above the laws of nature that apply to other animals. We talk of our 'descendants', implying a downward movement as though we ourselves have descended from the gods and not ascended from the animal kingdom. That we experience something we call 'free will' is no bar to our also being influenced by instincts. Lorenz then offers his attempt at an objective, dispassionate picture of our species as it might appear to an alien ethologist—one using Lorenz's methods, of course, subject to the limitation that the observer has more experience with simpler creatures than with this more complex animal. And he finally offers a cure or two for the malady of uncontrolled aggression in man, which again is a personal and limited prescription. Lorenz believes he is right, but even if not (says Tinbergen) he has shaken up accepted ideas in such a way that truth may emerge.

The facts of these chapters are not beyond challenge. In the case of the anthropological material quoted by Lorenz on the aggressiveness of the Ute Indians of the American prairies, there are opposing views. A second anthropologist who has lived with the Utes wrote a paper presenting a radically different view of them from the one accepted by Lorenz. The later evaluation removes the Utes' value as a case of man's instinctive aggressiveness. Michael Cullen, an ethologist of the Oxford School, sums up *On Aggression* as a work of wisdom rather than knowledge, though containing a message of profound significance for those who would study man.

If the aggressiveness of animals generally falls short of killing or even serious injury, what accounts for the lethal aggressiveness of war between men? For once Lorenz chose not to extrapolate from the animal example he had to hand—rats—but turned instead to other possible answers.

On an occasion when both shared the platform at a meeting in London, Lorenz heard a suggestion from Erik Erikson which he has quoted many time since: that man is capable of 'pseudo-speciation'. In the extreme case, only members of one's own cultural group are seen as truly human, while others may be regarded as belonging to some lower species that can be killed with impunity. Certainly there is evidence for this in the appeal of demagogues. In all animals (including man) in which there is strong cohesion within each group, this correlates with a capacity for aggressiveness between groups. Lorenz has seen this in his geese. When an alien group of geese is near, members of the home flock perform their greeting ceremonies of love and friendship more intensely than if they were alone, and this avowal of courage is a measure of their readiness to fight. So far, human and goose societies are alike. But man goes further: more easily than any other animal, he can release himself from the general rule against killing other members of his own species. In war, airmen may drop bombs to kill defenceless women and children far below without any feeling that the flyer's own humanity is prejudiced by his dispensation of death and mutilation. According to Lorenz's argument, he would manage this by declaring (and believing it to be true) that his opponents have forfeited their right to be called human, and may be killed as a lower pseudo-species—with hatred, insofar as they may still press their unwholesome claim to be men, or with casual indifference if that claim is not asserted or not seen.

But there may be other, and better, explanations. Tinbergen, among others, suggests that modern man derives from a mixture of two of the forms of society that Lorenz discusses in his book. Alongside the capacity for love and friendship at the personal level is the group loyalty that derives from man's evolution through a phase as a social hunting primate. As a result, man has developed characteristics in common with wolves or other pack-hunting animals, and this sets him apart from the primates that do not hunt but simply gather their food. (It also helps to explain the otherwise confusing fact that ape and monkey societies are generally very different from ours.)

Individuals, pairs and small groups all play their part in human society. If their aggressivity is misdirected into antisocial channels, there is a case for the application of social sanctions, perhaps formalised as the rule of law—and this in Tinbergen's view (of 1968) presents no further big social problem. At that time he could concentrate his concern on the effects of the potential misdirection of group aggressions on a much larger scale in the pursuit of war—the cases in which group loyalties are inflamed and misdirected at a national level.

Since 1968 we have seen more attempts to introduce a third type of social structure; the large family (or commune). This has not been very successful, as the conflict with pair formation often becomes acute. But its violent analogue, the urban guerilla or terrorist group, has achieved a considerable social significance. Such a group has internal loyalty and the potential for aggression towards other groups, qualities which at lower levels may be perfectly normal in human society, but which are seen as pathological if the intensity of their opposition to and action against 'enemy' groups goes beyond the limits set by society. In fact, it would now seem more appropriate to classify such smaller aggressive groups with the national groups whose conflicts are called war. It is notable that the members of such murderous groups talk of their 'campaigns', wear uniforms when circumstances permit, expect like-minded outsiders to admire them, and can be genuinely astonished when society does not respect their method or purpose. There does appear to be a significant difference between the aggression where an individual acts either for himself or on behalf of some other individual with whom he shares a direct bond, and the sort of aggression that is associated with loyalty to a group.

In his 1968 paper, Tinbergen discussed man's accelerating cultural evolution, the passing of experience from one generation to the next that produces changes incomparably faster than can be achieved by genetic evolution. Our increasing dominance over the natural controls which limit the activity of animals has created an explosive growth in population. We live too close together, he says, many of us well above a reasonable population density for our species. Such a high density leads to the continuous external provocation of aggression.

Lorenz agrees with most of this. Still evidently assuming his 'hydraulic model' of an instinctive pressure that if bottled up will eventually burst out with explosive force, he proposes that we should provide more opportunities for the redirection of aggression into harmless channels, such as sport. But Tinbergen, among many others, is not entirely happy with that particular solution, since for spectators as well as players, the aggression in sport can often be self-reinforcing rather than cathartic. He would prefer to offer the path of sublimation in creative activity and recommends especially the pursuit of science itself. He suggests that the public should be persuaded to participate (again, if only as spectators) in the solution of major problems facing mankind — such as that of aggression.

Peter Marler, an English ethologist at Rockefeller University, New York, offers more general solutions: to cultivate unresponsiveness to stimuli that might provoke aggression and to take consummate care in avoiding the provocation of others into aggressive activities. This could be applied both to individual and to group aggression. It is a radically different solution from that offered by Lorenz in that the underlying theory itself is changed: the hydraulic model is abandoned in favour of a system of graded responses in the interaction between individuals or groups. This is a more popular approach today, even to the extent that formulae to predict the outcome of aggressive encounters have been set up as mathematical inequalities calculated in terms of cost and benefit, and territorial boundaries defined by the solution of mathematical equations.

In discussing human 'territory', both Lorenz and Tinbergen seem to follow the animal analogy closely. The territory of animals is the physical space they dominate and hold from seizure by other members of their own species. For man, 'territory' could perhaps be more broadly defined. The territory that each person claims will depend upon his individual capacity and skills, and there is an almost limitless range available to a man who specialises enough. Specialist scientists (such as the ethologists themselves) form large international pseudo-tribal groups with a vast amount of space for each individual: it is not surprising therefore that such pursuits are deeply satisfying. Obviously, there are many less fortunate people whose occupations crowd them with their fellow workers as closely as they are crowded among their neighbours in simple physical space. By the standards I have defined this may appear as a social evil; but there are still vast opportunities for individual expression in private lives, and these are accepted and pursued by many who find them very rewarding. The social unit of the city, where crowding is greatest, also has the greatest opportunities for diversity and so for the existence of rare, small groups with common interests. Physical crowding may create tensions, but in some degree it must also help create the conditions for a vast diversity of opportunities for their shared sublimation.

In Hinde's view, too, a belief in the inevitability of aggression is unacceptable. Rather, we must look at the situations which precipitate violence and at the factors in the development of the individual which predispose him to aggression: we should examine the roles of early thwarting and achievement, experience of aggressive encounters, and frustration and crowding. These should be within our control.

Also according to Lorenz, society may suffer less from the results of misdirected individual aggression than it does from misdirected love, which can cause social disruption and permanent damage to individuals. And he sees man's over-competitiveness as another even greater vice: he compares it with the displays of those birds which are so heavily decorated with gorgeous tail feathers they can hardly fly. In contrast, he sees aggression within normal bounds as generally good: when we talk about 'tackling' or 'attacking' our problems, we envisage a positive use for the aggressiveness which man needs both for satisfaction as an individual and survival as a species.

Predictably, the most outspoken of his critics are to be found among the ranks of the behaviourists, together with some anthropologists such as Ashley Montagu. Lorenz sees the attacks on him as having the quality of religious fervour aroused in men who sense the collapse of long-established dogma. 'Ashley Montagu has formulated that dogma with all desirable clarity: man is devoid of instincts, all human activity is learned.' In fact Montagu does not go to this extreme, but he certainly does rate instinct—which he describes as an inborn predisposition to behaviour towards a preferred subject—very low in man. Indeed, he comments, we even have to learn our sexual behaviour. Hormones act on brain and body influencing our thoughts and behaviour towards the preferred sexual object, but we still have to learn what we must actually do.

On aggression itself, Montagu says that Lorenz is quite wrong in suggesting that prehistoric hordes automatically engaged in violence. The reverse in fact is true: first approaches between groups have generally been friendly. Montagu accepts that man has an appeasement mechanism—the human smile is understood where language is not—and that other expressive patterns such as those of rage or amusement are inborn. But he also believes that every activity beyond this level must be learned from other human beings. Montagu is mild and compassionate but somewhat rigid in his views. He has collected a range of articles and papers which dispute the views of Lorenz and Ardrey and published them under the title of *Man and Aggression*. The volume contains two papers opposing Lorenz on the Ute Indians, and a study by Geoffrey Gorer showing that Lorenz's apparent acceptance of the view of *Australopithecus* as an instinctive killer is unwise: the facts are far too flimsy to bear such an interpretation. (In fact, Lorenz does not claim that man has a killer instinct, but rather that in aggressive encounters in both animal and man there is a strong selection pressure towards mechanisms for sparing a victim.) Lorenz's evidence on 'bloody mass battles' between groups of rats is disputed and qualified. Above all, he is accused of largely disregarding previous findings on human aggression in his naive enthusiasm for drawing false parallels between animal and man.

But in sharp relief to all this, it should also be noted that among Lorenz's staunchest supporters is America's leading woman anthropologist, Margaret Mead. She supports him in his views on the domestication of civilised man as well as on the question of the aggressivity of man.

In Europe, attacks come from left-wing social critics, Helmut Nolte, for example: *On Aggression* 'has been hailed by the public as a form of absolutism because it comforts man that his behaviour in the context of present society reflects purely natural conditions.' In this the Left is in alliance with some of the critics collected in *Man and Aggression*. For these, Lorenz (and Ardrey even more) let man off the hook far too readily. They feel that the reader of these authors could all too easily extend their dangerous argument to the conclusion that violent crime and murderous wars are not our fault but a legitimate part of our equipment for survival—so, if our neighbour is likely to be aggressive, we should hit him first!

Nothing could be further from Lorenz's mind than this. He frequently stresses that man, with his capacity for reason and speech, is uniquely a great deal more than the sum of his instincts; but that even so far as he is guided by them, they are the very basis of his morality and ethics. Here Lorenz is most certainly being misunderstood. His position is that aggressive behaviour is inborn and can by the conditions of modern society easily become pathologically misdirected – but this does not mean that we have to leave things as they are.

Orthodox Freudian psychologists also reject Lorenz. For them the decisive roots of human behaviour lie in early childhood experience; aggression (or aggressive behaviour as most people understand it) is not inborn but the result of early frustrations. As we have seen, Robert Hinde accepts this as a viable hypothesis: but does Lorenz? At first he rejected Freudian psychology totally because Freud goes on to link his whole system to sexuality: that aggression derives from early *sexual* frustration Lorenz saw as obvious nonsense. But on a post-war visit to America he found an attitude to Freudian psychology that was entirely different from the rigid orthodoxy he was used to in Europe. The Americans were more open-minded, taking Freudian theory as a working hypothesis. They accepted those parts which seemed useful and disregarded what was not. Lorenz was happy to follow in their footsteps: he accepts the Freudian repression theory and rejects the sexual part. Lorenz's pupil Eibl-Eibesfeldt says that on infant sexuality Freud simply read the behaviour signals the wrong way round: the correct interpretation is that adult courtship in both animal and man includes some of the behaviour of infancy. Many birds, for example, include ritual feeding in courtship, and the human kiss is essentially the same action.

Irenäus Eibl-Eibesfeldt now heads a relatively new human ethology outstation of the Max Planck Institute for Behavioural Physiology based in a house on the outskirts of Starnberg, a brief *autobahn* drive south of Munich. He takes his inspiration very directly from Lorenz, whose photograph occupies a place of honour on his desk, and he regards Lorenz's indication that ethology should be applied to human affairs as one of his greatest contributions. Lorenz was instrumental in getting this working group set up; it was clearly a good way of encouraging the ethological work on man that is necessary to validate his hunches scientifically. Besides his ethological textbook and many papers, Eibl already has to his credit one popular best-seller, *Love and Hate*, which his friends have jokingly called *Die sogenannte Liebe*, so neatly do its statements and style fit with that of Lorenz's Sogenannte Böse, the so-called sin of aggression. In this book, Eibl-Eibesfeldt stresses the counterbalance that exists between man's aggressivity and his strong bonding impulses. Among his favourite stories is that of soldiers in war who are placed in trenches close to the enemy. When the opponents can openly see each other they will eventually stop shooting and exchange cigarettes: human aggressors must be distanced to make them fight willingly.

So far Eibl has applied his method most powerfully to the study of children living at an institute in Hannover who were born both blind and deaf. He established that in these children the facial expressions registering anger, fear, crying, laughing, tenderness and much more (quite complicated programmes in Eibl's view) were not dependent upon learning by either sight or sound. These expressions are certainly distorted in their appearance, since some feedback or example is evidently necessary to their final form, but there can be no question that they are present. These are, however, the least disputed (though certainly not undisputed) of pre-programmed human actions or reactions. Watching those deaf and blind children, a behaviourist might well point to the fact that their remaining senses are made to work overtime to absorb information from the world about them. Eibl himself could go on to draw attention to the way in which they draw back from strangers, displaying a shyness that is not warranted by experience, since strangers are always kind to these deprived children.

In a second tier to his work, Eibl-Eibesfeldt has made cross-cultural studies, wherever possible making a filmed record. His reflex camera has a mirror in front of the lens so that he can seem to point his camera in a different direction from the true subject. By this technique he aims to avoid any disturbance of their behaviour by reaction to the camera itself. He looks at the major cultures of the world, and for further comparison seeks out tribes that have had little contact with Western man. Unlike most anthropologists he is looking not for cultural differences, but mainly for those species-specific responses that all men have in common. Patterns of threat; patterns of submission such as pouting or looking away, are, he believes, inborn since they have been found in all cultures he has studied so far. He picked up one unexpected but intriguing detail from the frame by frame study of his films: this was a rapid, transient raising of the eyebrows at the moment of recognition and greeting. It is a pattern we may not be consciously aware of, and yet we transmit and receive the signal many times a day: this, too, appears to be innate. On the other hand, a pattern that must be culturally determined is head-shaking for 'no', for some cultures use different signals, thereby frustrating attempts by non-linguists to converse with them in sign language.

Part of Eibl's plan is to determine, if he can, just how much and in what form aggressivity is inborn in man. This is an essential first step towards controlling or redirecting those parts of human aggressivity which may be at odds with our society or culture. (Or, presumably, it could alternatively be a first step towards changing our culture to make aggressivity less dangerous.)

Otto Koenig, at his institute in Vienna, has sought out patterns that are instinctive in animals but have been adopted in human cultural practices. Only recently has man gone to war in anything but the most splendid uniforms he could afford or create. By so doing he wore the territorial feathers of his non-human ancestors — though, clearly in man this was also an expression of the group territoriality proposed by Tinbergen. The senior ranks favoured broad shoulders (epaulettes) and flamboyant headgear, sometimes with real plumage, in order to produce the imposing, enlarged outline that birds and beasts achieve when they ruffle their feathers or fur to intimidate an enemy.

But Lorenz's followers have so far only nibbled at the corners of human behaviour. The mainstream of orthodox psychologists still have the strongest hold on the subject; Leon Eisenberg is one representative of that mainstream. His attack on Lorenz was only in part directed at the 'Nazi' paper; rather more he disliked Lorenz's readiness to extrapolate from animal to man. Says Eisenberg, 'As far as I am concerned there is a quantum jump in moving from man's nearest primate relatives to man. That doesn't deny continuities; that doesn't deny the existence of physiological systems that are remarkably similar in different species, but I urge caution in extrapolation from animal to man.' Man is characterised by language. The single word 'fire' shouted in a crowded theatre can create an enormous physiological response, and yet it is merely a symbol that stands for an event. Such symbolic communication is not possible in other species. Using his capacity for conceptual thought and communication, man, in the course of his transactions with nature and with his fellow man, creates his own uniquely human nature. This is not present at birth, but is passed on to him.

So far I cannot see that he has very much to argue about with Lorenz. Lorenz agreed that man has moved beyond other animals to the extent that his has become almost a separate kingdom. From the virus upward to our nearest ancestor in the Olduvai Gorge, the only way of passing on considerable amounts of information was in the genome; it had to be coded in the nucleus of the cell. But now we suddenly have conceptual thought and speech and can transmit adaptive knowledge in vast quantities from one individual or generation to the next. This is nothing less than the inheritance of acquired characteristics. From knowledge gained within his own lifetime man can change his own response to nature and pass this change on to the next generation. Since it is Lorenz saying this, there is clearly no substantial dispute, for it is precisely what Eisenberg means by 'a quantum jump'. But Lorenz repeats that man's separation from the animal kingdom cannot be total: it would indeed be very curious if among all the structures of life just one, the human brain, a very complicated organ, was devoid of structure developed in the evolution of our species.

It is easy enough to criticise *On Aggression* on points of factual detail and on the opportunities it offers for misunderstanding. It is also possible to disagree with Lorenz's interpretation of human aggressivity or his suggestions for holding it in check. But there is no doubt that he succeeds in 'shaking things up', as Tinbergen put it. And if he is wrong? Another friend of Lorenz—one who did hold some of the possible reservations—said this of him: 'A man who has done so much has the right to make mistakes. And even so, he is pushing science forward.'

Indeed, in my view, this is what has happened, for ethologists (though not necessarily those who have stayed closest to Lorenz) are in the front line of current research on aggression, alongside psychologists and scientists from many other fields within biology. Few would still defend Lorenz's hydraulic model of aggression as more than a partial explanation using a convenient analogy that works in certain cases, or prescribe sport as its safety valve—though team games may yet remain as part of training in formalised group aggression, and for the development of each individual's awareness of this capacity within himself in a controlled situation; while for the spectator they are still desired as (at best) an enjoyable focus for group identification. But these are peripheral matters. The main point is that the present leading ideas on aggression do not at their core represent a rejection of Lorenz's theories, but by being based on the twin foundations of good observation and the best available current interpretation of Darwinian evolution, they stand closely in line with them, building on what is given.

Chapter 12

The Sins of Social Man

At sixty, Lorenz could still look out over Seewiesen's small lake and call in his geese — greylags, bar-heads, snow geese, and even a mallard or two, some eight or more different species in all, for comparative study. He also still swam with his geese, but was furious (in retrospect) when *Life Magazine* published the remarkable photograph by Nina Leen that was often used later to symbolise the man and his work. It shows two snow goslings symmetrically placed at either side of the grizzled head — all of Lorenz that showed above the water — as he grinned into the camera. This, he felt, anthropomorphised a random, natural, meaningless event; it devalued the intended demonstration and trivialised his science.

A few years later, in his mid-sixties, he could still look down in the middle of the day and recognise most of the older geese as individuals. But some he could not immediately identify, let alone the majority of the younger ones, for by then he had become somewhat remote from the work with which his fame had linked him, although maintaining an interest in studies of the social forces that hold such a group together. In one of his own last goose experiments, he did cast himself in a major role and played the part of one of the three goose 'mothers' who, by agreement, set up a rank order among themselves by making the appropriate signs of dominance and submission that would be recognised by the goslings. Each brood quickly learned which could legitimately bully the others, so that in some cases a smaller and weaker gander from a high-ranking brood could lord it over a bigger fellow from a family of lower rank.

But apart from such matters, he was at least equally interested in his fish, and most concerned with social or bonding mechanisms. Tropical fish also have certain advantages over geese: they are easier to breed, can do so all the year round, and some (such as the cichlids) even demonstrate elements of behaviour similar to those of his birds. Certainly, he has always been interested in the behaviour of individuals; but to overlay this there grew in him an increasingly urgent desire to understand as much as possible about the nature of society, the forces that hold it together, and those that may tear it apart.

Some scientists who saw his animal work during this phase of his life have criticised or commented upon the reduced flow of original papers on animal ethology arising from these later interests. It seemed almost that he continued animal watching through habit rather than in a spirit of positive or specific enquiry. But his mind was occupied with the social behaviour of man, a matter too complicated to observe in the laboratory. He was using the activities of the animal life before him to trigger trains of powerful analogy.

He could also ponder on the way he had become identified with one particular aspect of behaviour—aggression—to the exclusion of the many others that had interested him more in the past, and the broader social issues that concerned him now. In the introduction to his republished collection of older papers, *Studies in Animal and Human Behaviour*, he recorded that it was the stormy reaction to *On Aggression* which showed the need to lay out in one place the scientific path that he had followed. Clearly, this would not reach the wide readership of *On Aggression* but at least it would be available to the serious student of ethology and to his popularisers. From the sheer volume of words in the two weighty books it is difficult to pick out an argument in concise form, but this task was attempted by George Stade of Columbia University, New York (who also took into account his other books in English).

From internal evidence, and in the absence of any explicit summarising statement by Lorenz, Stade sought to define the political position of the writer of the papers before him. He recalled the criticisms from the left—that Lorenz's commentary on the modern world of man was merely 'the grumbling of a conservative, if not the bile of a reactionary or the ravings of a Fascist'—and concluded that there is simply no evidence for saying that Lorenz is a (political) conservative but rather that he is a radical, in so far as the lessons of ethology are themselves revolutionary. In the true political conservative Stade sees a tendency to believe in original sin: that we are naturally vicious and require good institutions to keep us from going to the devil. In contrast to this, the Left may take either of two views: that we are born without good or bad in us and are what society makes us; or that we are naturally good. To Lorenz, as we have seen, the 'blank sheet' at birth is behaviourist blasphemy: he ardently promotes the view that we are born with an apparatus for both moral and aesthetic judgment that is necessary to the preservation of our species. On this analysis, the leftist political attacks on Lorenz are all the more bitter because the transgressor is a non-conformist potential leader of leftist opinion, and therefore a closer threat than any common enemy.

Stade finds that in Lorenz's scattered remarks on social and political issues there is a pragmatic liberalism: 'He stands in that shifting and uncircumscribed zone in which left-liberalism, social democracy, and democratic socialism jostle one another.' He is for the rights of women (within ethological reason). He is against nationalism, imperialism and racism. He is pacific and a pacifist, and he is for many of the causes espoused by the militant young (if, by membership of a different generation, he must be against their militancy). The one significant element missing from Stade's reading was the untranslated papers of 1940 and 1943. Lacking the explicit political jargon of the first, and the hints that would have been supplied by the second, Stade comments that while the sum of Lorenz's individual statements finally defines no particular position in modern politics, it disqualifies him for Fascism. That is a verdict with which the Nazis evidently agreed.

When his son Thomas told me with emphatic certainty, 'He's not a conservative, he's a revolutionary!' I suggested that if this is true, it would seem that he seeks the quietest of revolutions. While he may be no conservative, he is a conservationist; he is prepared to retain the workable parts of any existing system, adapting it by the minimum necessary change to a form that, in his view, will better achieve the survival of man and his society. Thomas translated that biological statement into the language of physical chemistry: 'Minimum entropy—yes!' Then he thought for a while and added, 'Low entropy: you can see it even as early as the 1940 paper'. Some such signs may have been there so early, but if one had to choose a date at which it could be said that the revolutionary had become an evolutionary it would probably be when he was appointed Director at Seewiesen.

In his social comment, the points at which Konrad Lorenz condemns also fall into a pattern. He seems eventually to have decided that the learning-by-discovery style of *On Aggression* was too indefinite, and that the literally voluminous *Studies in Animal and Human Behaviour* required a slender sequel to summarise his conclusions on Man. Thus *Civilised Man's Eight Deadly Sins* appeared in English in 1974 (following the German edition of 1973) and yet by the time it came out he had already modified its pessimistic view. Because continental Europe

generally was later than America and Britain to reflect the environmentalists' demands for action, Lorenz was ahead of local opinion
when he wrote the book; but he found himself far from being in the
lead (giving the appearance of jumping on a bandwagon) when his
strictures took the reverse route—westward—from that followed by
environmental activism.

In his comments on man he can be accused of daring too much too
incautiously by comparison with his discussions of animals; and, as
he grew older, of losing some of his capacity for self-criticism, while
retaining that for crisply phrased assertion. It is better in the circumstances for the more critical reader to see *Eight Deadly Sins* as a set of
arguments for which most of the foundation lies in Lorenz's earlier
papers and lectures—which does not prove him right but merely
shows that he can back up most of his statements with supporting
examples from the animal kingdom, and has reasons why each analogy
should offer a reasonable model for and interpretation of the behaviour
of man.

At a second level, more acceptable and possibly more convincing to
some, he would appear to have given up such round-about arguments,
turning instead to direct lectures on the human condition and its social
pathology. The puzzles and paradoxes of man's irrational history are
made understandable by a view of man in which his behaviour is still
partly governed by the rule of instinct. He tells us that we need the
motivational system this provides, but the system is finely balanced
and civilised man may all too easily step away from the narrow pathway on which that balance can be maintained.

Perhaps the most interesting discussion is that surrounding the sixth
sin which concerns the 'generation gap'. The basic thesis is one that
Tinbergen and others have also argued, that the growing rate of
technological advance creates increasingly different societies from one
generation to the next. Inevitably then, the difficulties of adjustment
grow with this. In the supporting argument, Lorenz carries forward
some ideas already seen in *On Aggression* and developed on a number
of occasions since then. In doing so, he makes clear not only the
analogy he is using, but also, once again, that he is well aware of the
vast differences between human and animal societies.

The basic analogy he uses here is that between the evolution of
animal species and the evolution of human culture. For any animal
to change either its form or its fixed pattern of instinctive action, it
must change fundamentally in the storehouse of its preserved informa-

tion for future generations, i.e. in the genes of each animal of the species. Animals can and do have closely circumscribed capacities for cultural tradition, but for the most part mutation and natural selection is the main if extremely slow way that a species adapts to new conditions. If the environment of a human being changes, he works out what to do about it—and the next generation knows at once. As we have seen, in human cultures we do not have to wait for mutation; we transmit the acquired characteristic directly. Each new method of response achieves in time the status of a tradition, which may take the form of superstition, myth, doctrine, or rite, or may be codified as law or taught as recognised academic knowledge. In the span of human cultures, these external bodies of formalised information form a second tier that overlays the message of our genes. This capacity to entrench tradition is a vital property of the apparatus that man has developed for the preservation of culture. But we must be aware that in such a system it is risky to remove elements arbitrarily, even those that are apparently bad, for they are part of a coherent system of a complexity comparable to that of our instinctive behaviour patterns. They are so intricately interlinked that pulling out one brick may topple the entire structure. Anthropologists rightly warn against subjecting primitive tribes to 'culture shock'. A culture is not easily re-directed from without, but can be all too easily destroyed—and the humanity of man, deprived of its supporting culture, is destroyed with it.

Cultural traditions or doctrines define a society in the same way as its genetic constitution defines a species of animals: each culture forms what may be seen as a pseudospecies, regarding its own traditions and doctrines as part of the fullest definition of what may be taken as truly human. Lorenz warns that it would be arrogant of scientists to suppose that, when science provides a simple and apparently sound solution to human problems, it can be introduced to the societies of the world without extreme care. People who deal with overpopulation and undernourishment throughout the world are by now aware that when the technical and economic problems have been overcome, cultural obstacles remain. To crush the cultural resistance is as arrogant as interfering with human genes.

An extension of Lorenz's argument is that all revolutions must, by their very nature, fail. The trauma of failure is in direct proportion to the strength and quality of the cultural tradition that must be torn down in order to build anew. Human adaptability is great enough for the new growth to be strong and fast: revolutionaries rely on this, but change as violent as that in the French or Russian revolutions produces long-lasting shock and temporary systems of government (or culture) in which much that is human may be debased, until a new culture that is only in part the creation of the revolutionaries grows up. To the extent that Lorenz must oppose any form of revolutionary change as ethologically unsound he is, after all, a conservative, and it would be difficult for any biologist not to follow him. To the extent that he proposes change by cultural adaptation, a little at a time, but as much (and as little) as is made necessary by the problems we have created, Lorenz can be a moderate radical. There too it is easy enough to follow a similar path, though not necessarily the same one, since each of us may disagree about what is most urgent in the short term. It is a task for the most skilful politicians to plot the least disruptive course of positive action that will effectively deal with the growing problems that we face.

In his chapter on the generation gap, 'The Break with Tradition', Lorenz introduces another element in the mechanism for the evolution of human culture: he suggests that human culture has adapted a form of behaviour that in his geese (and many other animals) is used for another purpose. As geese grow up they reach a point where they seek their own companions, rejecting to some extent the company of the parents; this is necessary to prevent inbreeding. In humans this inhibition remains, although the capacity of humans to override a simple instinctive inhibition has required the addition of cultural sanctions, in this case both religious and legal, in a wide range of otherwise diverse cultures. Man's capacity for cultural adaptation arises largely as a by-product of the assertion of independence that already comes naturally to the young. It occurs at an age when the cultural heritage offered to the child can be examined against the future needs of the society he is to inherit. He will seek to adapt it, the better to serve his own and future generations.

Such a mechanism, if it exists, must itself be a genetically deter-mined element of human behaviour. So it follows that the controlling mechanism cannot itself adapt as rapidly as the change in human culture that it permits, but could continue to function in inappropriate circumstances. The danger point comes when the degree of techno-logical or any other sort of change between one generation and the next is such that a fully adapted individual might have constructed a culture sufficiently different from that of his parents for each to see the other almost as a different species. Such a gulf, argues Lorenz, could elevate the natural rebelliousness of the young to the level of enmity between pseudo-species.

Perhaps this has already happened? Yes, says Lorenz, when it can be illustrated by cases where cultural instability has given rise to adolescent subcultures, often with a distinguishing manner of dress for cultural identification; yes, where the young are driven to join any culture at all rather than belong to none — even though it may be a drug cult that could lead to death; and yes too, in the case of much student protest where the young assert independence by indiscriminate cultural damage and deliberate offence against received standards; but no, when the issue finally selected is one of those where a profound change within the time span of one generation is the only way of safeguarding humanity from disasters that may themselves amount to a self-inflicted cultural genocide.

And he has given reasons for hope: firstly in the very ubiquity of youthful protest, whether it be against Stalinist orthodoxy in Eastern Europe, the antiquated professorial tyrannies in German-speaking universities, or the complacencies of the 'American Way of Life' in the United States; and secondly, in the fact that youth does not demand change 'in the wrong direction'. He tells us, 'Never have they demanded a more effective commercial system, better armament, or a more nationalistic attitude of their government'.

But such statements of his optimism have been less pronounced — or less quoted — than his criticism, of which it could be suggested that he has sometimes gone too far: that he has adopted the defensive attitude of counter-attack that the protest is designed to provoke in order to achieve conflict. Lorenz is certainly aware of the danger of this and tells us that he has seen his fellow professors make fools of themselves by fighting back, or heard them branded as cowards when they have given in to every demand — with the strongest abuse reserved for those whose political opinions lie closest to those of their young opponents. It is not a prospect he can view without emotion, and even while he analyses it he can hardly avoid some of the traps that are set for him. The perversion of a necessary mechanism can loom so alarmingly close to the eyes of an academic and teacher that he may overestimate the incidence in the population as a whole. If, at the peak of student unrest, active protestors comprised even ten or twenty per cent of the student population (a generous estimate), this may still be only two per cent of all youth; while no more than the tiniest minority join gangs or groups that are truly pathological sub-cultures. But in assessing even the most extreme cases we are faced by real problems of judgment, for in groups that many might regard as 'cancers of society', fine human feelings may still be expressed among members. For Lorenz, this was demonstrated movingly, poetically, and with impeccable ethological accuracy, in the musical, *West Side Story*, where two rival adolescent gangs form sub-cultures that bring out in their members the good characteristics of man—loyalty, friendship, selflessness and many others—to serve the empty cause of a murderous enmity.

At a 1970 symposium on 'Play and Development' that included Jean Piaget, René Spitz and Erik Erikson, Lorenz's contribution was forcefully entitled 'The Enmity Between Generations and its Probable Ethological Causes'. The title apart, this paper handles its subject better than does his popular book, in which the biology is almost an intrusion among the assertions. When talking to his scientific peers he is no less assertive, but he tempers his argument with more biological support, and the ideas so expressed—good reading in their own right—contain the matured results of his earlier thinking and allow the reader a better chance of judging his castigation of the young in the light of his own science.

It is easy enough to argue that he grossly overestimates the magnitude and importance of any recent rejection of existing high cultural standards—which has in any case died down among students in recent years—but it would be difficult to counter his statement that the alienation has actually increased in fifty years of his own lifetime. Given the increasing rate of technological change, it would be surprising if the adaptive mechanism had not moved into a higher gear. Such a shift is bound to give greater reinforcement to its pathological forms, which will inevitably be more blatantly expressed. In both book and paper Lorenz seems in the end to be in two minds. The truly reflective young, he says, are the less violent ones; and he has found these even among young people who have expressed strong opinions (like himself?). In his youth he too fought strongly against his father in the matter of his own education, and if he lost it was on his own terms; on the question of his betrothal he won. The early growth of his science depended strongly on a healthy disrespect for established but misguided authority. Lorenz knows perfectly well that progress is achieved not by the eventual conformity of the young to a totally unchanged tradition, but by a process of mutual adjustment, in which the finally adopted values are not necessarily those of either group. He insists that the danger remains: it is virtually impossible to convince the young how important it is that some substantial proportion of cultural traditions be conserved; that a culture can all too easily be snuffed out. But he accepts that blame for the excesses of the late sixties and early seventies might have to be shared by older generations who failed to let the young participate sufficiently in the ideals for which all should be fighting. It was noticeable (he adds) that in biology the student troubles were minimal, while in sociology, psychology and politics they were tremendous: we are left to draw our own conclusion about which science has offered the best opportunities for the identification of problems and has opened up avenues towards their solution.

Indeed, a striking change had occurred between his writing the manuscript of *Civilised Man's Eight Deadly Sins*, and its publication three years later. At the time he had felt like a prophet crying in the wilderness, but when it appeared in print he became at once the author of a German-language best seller. He recalled the 'martyrdom' of Rachel Carson who 'died more or less discredited, hunted down by the industry', but who had now become, in effect, 'the patron saint of millions'. The public, he concluded, is after all more receptive and more intelligent than the mass media sometimes give it credit for — but does he (like Rachel Carson before him) paint too pessimistic a picture? 'Look', he says, 'if you want to get people interested in their danger you have to make them somewhat afraid. If I paint the dangers very black, it is in the hope of getting people moving'.

When he wrote the book he was right to be pessimistic: ecology, for example, was a virtually unknown science and very few paid attention to important ecological facts. Clearly, that is no longer true. There is also what, at first sight, may be seen as a sharp reversal of opinion: the threat of nuclear aggression is not only last but also the least on Lorenz's danger list. Instead, he appears to be itemising a number of separate points which environmentalists and social critics have been making for years. We are overpopulated . . . destroying our environment . . . over-competitive . . . seekers of instant gratification . . . placing ourselves in danger of genetic decay . . . and increasingly in danger of manipulation and indoctrination by behaviourist techniques. The last two are statements in a new context of themes that he himself has long argued. The remainder are topics in which others have led public opinion, and Lorenz's contribution had been to see them not separately but as problems that are linked by and interlinked with man's genetic and cultural heritage and his capacities for evolution and adaptation.

An ethologist observing the behaviour of any individual within a species asks first the question 'what for?' meaning 'what is the survival value of this action for the individual or his group?' In animals, there are cases enough of individual behaviour patterns gone awry, such is the complexity of genetic structure and the animals' dependence on a suitable environment for the proper expression of innate patterns. But when he asks the same question of human actions, the answer is complicated by the overlay of cultural adaptability which promotes a great diversity of biologically 'permitted' behaviour. Man's control of his environment allows him to indulge in 'luxury' forms of behaviour which bear no clear relationship to survival; and there is also a much more complex behavioural pathology which follows from man's freedom to misapply previously valuable forms of behaviour in ways that, in our present civilisation, make them into social ills. Fortunately, in this last case there is some trade-off, for the pathology of a system usually makes the system itself easier to understand, and the observer can often relate the result to the over- or under-functioning of the determinants of a form of behaviour that is not in itself bad.

When balanced, nature has many systems interacting in such a way that the effects spreading outward from each individual system are subject to feedback from the others. Lorenz does not make a point of this, but there is no 'law' of nature which says that complex systems in the absence of man will be stable: indeed, the biology of the polar regions, which is relatively simple and subjected to the maximum annual climatic stress, seems more stable than that of the tropics, where the weather is much the same the year round, and life forms luxuriate. In nature, there can be gross natural eruptions (unexpected population explosions) in a variety of time scales, and the natural 'invention' of cultural evolution which has brought man on the scene has been accompanied by the most rapid and far-reaching of such natural 'eruptions' since life began on earth. Life itself is a system for creating order from disorder: unconstrained, living systems gather energy voraciously and grow exponentially. In practice, there always are constraints. The human life form has been passing through a period in which these constraints were weak but, in the years since Lorenz wrote his book, it has become apparent that we may be approaching one in which they are severe, and approaching it with a speed that will test our cultural adaptability in a manner that is totally unprecedented.

Our civilisations have established themselves during a rare ten-thousand-year moment of warmer weather in the earth's icy history, and at the most critical phase of population growth we face the prospect of a climatic downturn. If we can overcome the effect of fluctuating cooler conditions on our attempts to feed ourselves, and avert a consequent cataclysm of wars of desperation, there looms beyond that the threat of descent into a renewed deep ice age—a danger that was until recently unknown to science. It is clear that the casual shift of such a vast natural system could lay upon man stresses that are incomparably greater than the puny, misbegotten effects that he himself has so far had on his natural environment. In the past, the ice ages may even have given man an edge on other living creatures, for the earth's recent two-million-year era of drastic climatic changes must have favoured the evolution of such a rapidly adaptable creature. But now mankind has a new combination of stresses to face; from the viewpoint of a detached student of evolution, a fascinating prospect lies ahead.

But here I must apologise to Lorenz, for in attempting to simplify and extend his argument I have run ahead in a way that is acceptable to me but may be less so to him. His thesis on overpopulation links it to the ethologist's concept of territoriality: like predatory animals man requires physical space. Among animals Lorenz sees a good example of this in the comparison between Otto Koenig's two colonies of cattle egrets at the Wilhelminenberg in Vienna. Both groups have ready sources of food and nest-building materials; the main difference is that in the captive colony there is overcrowding within a confined space so that each individual's territory is reduced below the natural limit. The result is an almost total miscarriage of their normal social (and particularly sexual) behaviour. The loss of incest taboo and a reduction in the natural heterosexual distinction leads to indiscriminate copulation and group sexual activities that are not seen in the free colony. It can be seen here that the way in which innate elements of behaviour combine may be influenced by the single environmental factor of the living space allowed to each animal, a general conclusion, says Lorenz, which is also likely to apply to our own behaviour. Aggression levels are raised by crowding—but if the innate aggressive elements are carefully channelled we can avoid the violent outcome and still lead them into positive behaviour.

Man, says Lorenz, has a limited capacity for social contact. In the country you find real hospitality because this is not overstrained; in the crush of the city we can offer it to only a small selection of our neighbours: we are not so constituted that we can love all mankind. But the necessity to avoid involvement in the suffering of all our fellow men already conflicts with some of the principal qualities that culturally we regard as 'human'.

In fact, Lorenz in his book devotes relatively little space to over-population and perhaps, as a result, does not sufficiently take into account the effect of differences between animal and man. Man's individual adaptability allows him, the supremely unspecialised animal, to become specialised to almost any degree and in any direction. The result is a new kind of 'territory'. A city prospers by creating the conditions for vast numbers of village-sized groups of like interest within which the highly specialised can find fulfilment. Koenig rightly compares his captive egrets not to a city population but to a group of men doing a boring and repetitive job; and yet, demeaning as such a job may be, there remains in a free society with substantial leisure time opportunity for individuality in other phases of the production-line factory worker's life. Diversity creates territory, for each sphere of human activity has its own distinct and separate set of frontiers. A tailor is not in territorial competition with a doctor, nor a car assembly worker with an amateur ballroom dancer: all have their own competitive worlds while their territories outside their own interests may overlap without limit. Indeed, given sufficient specialisation, any number of men may each claim 'the whole world' almost undisputed.

There is no special difficulty in achieving this: it depends only on man's apparently inexhaustible talent for diversity together with a capacity to distinguish fine differences as defining separate fields of interest. We have already seen that even the statement of an opinion is analogous to claiming territory: a scientist may defend his theory to the last breath, and if in the end he loses we may say 'it broke him', or even 'it killed him', with real truth. Clearly, this argument could be pursued in such a way as to seem to attempt the refutation of Lorenz's position by an extrapolation of his own argument. This is not my intention. Rather, I would show that the malady (which does exist) contains the seeds of its own cure.

In his own 'black' picture, Lorenz sees man's devastation of his environment at least partly in terms of the effect that it has on man himself: his alienation from nature leads to atrophy of his ethical and aesthetic feelings. When all around him is man-made and tawdry, man himself is diminished. The urban periphery (unlike the older city centre) spreads like a cancer — and as the tumour cells have lost part of the genetic information that served to regulate their growth and relate them to their living surroundings, so the battery houses smother and reduce the individuality of their occupants to an ant-like interchangeability. This in itself represents a loss of essential cultural information. And all this, says Lorenz, is because modern man prefers action to reflection, and by his unconsidered deeds destroys the natural and cultural surroundings that have made him what he is.

And yet (or so it could be argued against Lorenz's position), nature has so far proved remarkably resilient against the worst insults that man has imposed upon it; whereas we in our present numbers may not survive one of the regular great cyclic events of nature. In the long run, man may see no alternative but to attempt the control of his environment to the extent of preventing natural changes that may crush his civilisation between ice and desert. Paradoxically, man's unintended experiments with nature may give him much-needed information on how its present form could be preserved — which lends a deeper perspective against which we may see our short-term problems.

Lorenz might argue that civilisation as it stands could never make it to the next ice age, soon though that may be in the time scale of civilised man: the false values of money-motivated competitiveness and the effects of luxury may destroy Western man first. From the ethologist's point of view, competition based on capitalist values is simply irrelevant; the battle for financial survival against other members of the same species does nothing to promote the evolution of that species in any direction of worth. There is nothing to promote goodness and kindness unless it is our own innate feeling for them. If commerce is against such qualities, it is fortunate that commercial fortune does not correlate with breeding success!

In all economically advanced societies we are cushioned from the success or failure that previously stimulated our development. Primitive man kept out of the way of all avoidable danger without a thought for the stigma of cowardice; but the instinct to avoid danger today may malfunction and express itself as a desire to avoid even the slightest discomfort. With the avoidance of unpleasure, therefore, the pleasure that depends upon the effect of contrast is also lost; our feelings level off in an artificial plain of emotional boredom. To escape this we seek ever stronger stimuli but ruin ourselves by the instant gratification of our desires. In sexual matters this is disastrous, for the gradual pursuit of distant goals in a slow courtship is necessary to strong pair-bonding. Instant copulation is rare in other high animal species, and is present only in domestic animals where man has bred out the highly differentiated mating patterns of their wild forebears in the interests of easy stock breeding . . . which leads Lorenz once again to his central theme of our potential genetic decay (of which the increasing juvenile crime rate is a symptom) and so to the problem, already discussed, of the gaping gulf between generations.

No one need agree with Lorenz on these dark fears, but it is surely preferable to have a sound argument against his view than to rely on mere hope that he is wrong; and the best arguments seem to demand a deeper biological understanding of man than we now have.

Lorenz's seventh theme of man's 'indoctrinability' shows him at his most discursive in an otherwise concise book. Several related arguments appear, and these go over some of the methods of research available to the biologist and discuss their validity. It is no surprise to see the behaviourist method once again condemned, not only for its limitations but also because a belief in the validity of this approach has led to increasing attempts to manipulate man by behaviourist techniques. To the extent that the method is successful—and there is no doubt it can be very effective—its use on man is the sin.

To Lorenz, the political implications of Behaviourism are 'horribly clear'. The idea that man is exclusively the creature of his conditioned responses, that he is malleable in every way so that if you catch him young enough you can make anything of him—all this, he says, is welcome news to anyone whose aim it is to manipulate man. Belief in the all-effectiveness of conditioning has moved on from being a scientific hypothesis to become a doctrine; even, in Lorenz's sharply critical view, a world religion. B. F. Skinner has proposed that the method is a power for good, that the modification of behaviour in man is a proper extension of the laboratory work with pigeons: 'If you discover how behaviour is related to the environment you can use the environment to predict behaviour or to control it. We've done this with psychotics, with retardates, with juvenile delinquents, with students in the classroom, and so on. You simply arrange a better environment and you get a better behaviour.'

Put this way there seems little objection to it—but operant conditioning used positively goes far beyond simply arranging a better environment. A system of rewards and punishments is applied to 'wanted' and 'unwanted' behaviour—and the decision as to what is wanted or not is in the hands of the person in charge of reward or punishment. Lorenz responds angrily to such an arrogation of power: 'The assumption that anyone—any human being—knows enough to take the responsibility to form and manipulate mankind is, in my opinion, blasphemy.' It is nothing short of an attempt to adapt human behaviour—and here he uses the term 'adapt' in the scientific sense of causing evolution under the pressure of environmental stress. He does not see how behaviourists can hope successfully to adapt something of which they know neither the structure nor the survival function.

In reply, we might ask whether a better understanding of human behaviour by the methods of ethology might not offer an even stronger weapon to the manipulators. Advertisers have been known to listen to talk of motivational analysis as well as to employ the simpler conditioning methods. Lorenz sees this in both East and West: in the East the barrage of commercial advertising that attempts to condition us is replaced by a monotonously insistent political message; the red posters of the eastern 'Big Brother' and the commercial television jingle are two sides of the same coin. Given such a desire to influence man, there seems little doubt that greater understanding of human motivation would find people ready to use it for their own purposes.

To this Lorenz replies that ethologists do not claim to know everything nor seek to use their knowledge to manipulate man, whereas the behaviourists do both. Skinner agrees that conditioning, like Otto Hahn's nuclear fission, can be misused, and there is good cause for alarm in the improper use of either. But it is a simple fact that they both work—and powerfully—and he concluded that the possibility of misuse does not disqualify either from being used for good. It is certainly true that many ethologists would like to see their knowledge used positively—to help backward children, for example. There is no doubt also that any technique is fraught with difficulty, but in principle ethology should have the edge since it seeks to draw human behaviour along channels which may already exist to some degree (because they do in most people), while the conditioning method assumes the construction of every item of behaviour from some external store.

Lorenz and Skinner see different aspects of the same real world: Skinner wants to modify while Lorenz wants to investigate the limits of modifiability. But this difference of interest is sufficient to place them subjectively in different worlds, with different political implications. Lorenz sees each individual as someone with his own inborn capacities and potential: equality is not inborn but it is reasonable and possible to demand as a political objective that each child should have equality of opportunity—an opportunity to express this potential and not to be subjected to levelling to fit a common mould.

The behaviourist assumption is that all people born with normal human capacities start out equal in life, for all practical purposes. Neither Skinner nor Ashley Montagu say that man has no innate behaviour; only that it is relatively unimportant compared with that which is handed to him culturally. Nor does Skinner deny man's animal ancestry, but he says that man has taken a different turn in his evolution—towards intelligence and the modification of behaviour in his own lifetime, rather than the building in of behaviours that are available to him at birth. What in scientific terms is really only a difference of emphasis between Skinner and Lorenz becomes more than this when practical questions of educational method are under discussion. To Lorenz, 'all men are born equal' is an obvious fallacy – and sound social policies cannot be founded on a lie.

One Lorenzian characteristic that emerges from any comparison such as this is that he does try to see man as a whole and to relate the problems facing mankind to the capacities of the whole man. The individual statements that emerge may be controversial—and a severe deficiency of his approach is that they are difficult to prove or disprove in rigorously scientific terms. Are we justified in calling such a discussion scientific at all? To attempt to answer this we must turn to the Lorenz least known to the general public, Lorenz the philosopher of science who seeks to evaluate scientific method in the light of his understanding of our human powers of perception.

Chapter 13

Truth in the Round

The central tool of science—as so many of us learned at school—is measurement: give the scientist something to count and he is happy. The most familiar laws of science can be expressed in mathematical form. The artist or poet is literate, the scientist is numerate.

Lorenz's science, however, is resolutely descriptive; and worse still, in this numerically encompassed world, he is often describing *qualities*. Many modern scientists might ask how Lorenz or anyone else can truly come to grips with an animal behaviour pattern or a human emotion if he is unable or unwilling to measure distance, duration, energy, angle or decibel—and not just once but many times over, so that some average can be discovered and the statistical spread on either side probed for meaning. Two observers who see the same complex event may describe it in very different ways—indeed, they are likely to do so—while the same two observers making a series of measurements are using a common language; their work is convincing because it is repeatable. How, then, can you be truly scientific unless you seek to quantify qualities?

For reply Lorenz tells a story against himself. 'How much do you love me', he once asked his wife, expecting a suitably romantic answer. She thought for a moment, adding it all up in her mind, then told him, 'eight'. Delighted by a response that so neatly exposed the question's fatuity, Lorenz passed it on to friends. 'I asked her how much she loved me and my wife said "nine" '. Gretl broke in, correcting him, 'No, I said "eight" '. Her husband should not promote himself!

To Lorenz there is a great deal more to this than a play on word and number. I asked him how many graphs he had drawn in his lifetime. He eyed me for a moment with the half-pleased, half-shamefaced air of a schoolboy who has decided to shock the world with a statement that defies all conventional belief. 'Do you know my story?' he began slowly, then suddenly he roared, 'I have never drawn a graph in my lifetime and I'm damned proud of it!' More quietly he added, 'I've done some very intelligent experiments, but on the whole I was the describer and my friend Niko Tinbergen was the experimenter'.

Tinbergen puts in cheerfully, 'I can draw a histogram'. (The occasion for this comment was a post-Nobel Prize discussion with physical and chemical fellow prize-winners on Swedish television.) Lorenz considered the marginal acceptability of histograms: 'I'll go as far as that, but not much further. You know, we are the whole-organism people.'

Their study of whole organisms places them at the top level in a hierarchy of complexities in all the sciences. At the next level down, there is the physiology of individual behavioural mechanisms, an example quoted by Lorenz being a study by the physiologist, Schwartzkopf, of how a grasshopper 'understands' and reacts differently to a rival's chant and the little clicks of the female. Schwartzkopf studied membrane potentials, tracing the information flowing in electro-chemical form from one cell to another in the ganglion chain, and showed how the sound registered by the tympanum is processed as it travels on its route to the motor centre in such a manner as to exclude all except the fighting reaction to the one sound and the courting response to the other. In this case, the phenomenon described by Lorenz and Tinbergen in a general way as an 'innate releasing mechanism' was analysed in terms of cellular activity, chemistry and the shift of electrons. In order to make progress, Schwartzkopf found it necessary to reduce the problem by narrowing his field of attack.

This is the way of scientific reduction: the whole animal can be analysed in terms of the physiology of its component organs; the physiology is further analysed down to its biochemistry; biochemistry can be taken apart to examine the physical nature of the chemical bonds and reactions, which in turn depends on the physics of the elementary particles of which all matter is composed. With each step downward in complexity, the scientist moves on to firmer ground, the laws seemingly become simpler, more general—and more mathematical. But this, at least in part, is a matter of how advanced each science is—or, from Lorenz's viewpoint, was at the point where he stepped in to give a new impetus to ethology.

The stages of scientific advance are observation and description, classification, the proposal of relationships between components of the observed system, the testing of the hypothesis by attempting to prove it wrong and then, if attempts at disproof fail, the provisional acceptance of the relationship as a scientific law. By its nature—with the simplicity and general application of many of the relationships that have been investigated—physics (and much of chemistry) at first advanced far beyond biochemistry; but in recent years there have been spectacular advances in that field too. One example occurred when a physicist turned his talents to bear on the genetic code embodied in DNA and helped to solve it as a problem in physics, chemistry and geometry.

Most biological problems are a great deal more complicated and so come later in time to the expression of general laws of a type recognisable by the mathematical physicist. The more complex the biology, the more 'primitive' the science appears, the more dependent it is upon the observation and description of a vast amount of detail, and the shakier is the translation of this into figures and therefore laws of a physical type.

Lorenz feels that biology—the study of life itself and hence the most important of our sciences—is consequently the most undervalued; to which it could be answered that in the second half of the twentieth century this is certainly not true for biology as a whole, which has moved centre-stage in the importance attached to it and in the interest and effort it attracts (although many current physical problems are more expensive to investigate and therefore may obtain more money for individual studies). It could also be true that compared with others Lorenz's corner of this large field is still rated low because of the difficulty of quantification. Lorenz feels that his colleagues give way all too easily to the pressure of current scientific orthodoxy: they are too ready to ape the style of today's physicists when they would do better to look to the pioneer. He points out that Isaac Newton was a mathematical physicist whose ideas on gravitation went far beyond any rigorous tests then available to him. It is certainly not true that his law of gravitation had little or no value because it could not be proved in the laboratory.

Today's physicist must rely more and more on instruments—his direct senses can tell him nothing about the movement of an electron or the spin of a pulsating neutron star—and it is right that such detectors should be calibrated: this adds a dimension to our knowledge. Indirect methods can be used in biology too, but their need is less to start with because what we are observing are living things which make signals, many of which are readily accessible to our own human senses. And yet it seems we have reached a stage where it is respectable for a scientist to exclude from practical consideration the direct evidence of his own eyes and ears. In fighting what he feels to be a rearguard action for the scientific method that he himself has used with such notable success, Lorenz has sometimes seemed to condemn all quantification in biology—although this, as usual, is an oversimplification of his true position, deriving once again from the assertiveness of his style. But with an effective force and humour he has been concerned to attack an aspect of this, 'the fashionable fallacy of dispensing with description'.

In contrast to the biologist, a physicist may more reasonably ignore the differences between a wrist watch and a grandfather clock and proceed directly to the relationship between the movement of the hour and minute hands. He may observe that one goes round twelve times faster than the other, call this 'the law of the clock', and then dismiss the differences in shape and size as beneath his notice. But the biologist is far less able to dispense with description, as variations in structure are as important to him as are the common features. In the final resort, it is differences of structure that make a man a man and a goose a goose, even though some of the functions of component structures may be the same. It is clearly essential for the biologist not to follow the physicist if this implies gross simplification. Some biologists do investigate only those features common to pigeon, rat and man: an example that we have seen is the behaviourist interest in reinforcement or conditioning mechanisms where all that is characteristic of the individual species (be it pigeon, rat or man) is deliberately set aside.

Apart from all this, Lorenz sees that a tremendous amount of descriptive work remains to be done; vast ranges of living creatures have yet to be studied in detail, work which cannot be replaced by experiment. Lorenz fears that necessary description may well be stunted by 'the craze for quantification'. Today, he asserts, you can hardly have a doctoral thesis approved unless it contains a plethora of graphs, statistics and complex mathematics; it is almost impossible to receive a doctor's degree for purely descriptive work. 'This is how far we are advanced in this fashion—which *is* a fashion', he insists. It prevents many young research workers from becoming true experts in their field. Tinbergen's view is equally explicit: 'Contempt for simple observation is a lethal trait in any science, and certainly a science as young as ours.' He argues that a balanced development of ethology is necessary.

Graduate students joining Lorenzian ethologists have often to go through an uncomfortable process of un-training. Tinbergen's former pupil, Anne Rasa, working with Lorenz at Seewiesen in 1973, found that students who came to work with her nearly always placed the stress on action: they wanted to know what they should *do*. With a new animal, she told them, do nothing; just look at it for a few weeks. Read no papers on the animal because in that way others' pre-set ideas are reinforced and important features not already noted may easily be missed. The students find this strange at first, but eventually accept the necessity to get a feel for an animal before they start counting things: long, intelligent observation breeds true knowledge. Nevertheless, Anne Rasa (following Tinbergen) sees the day of the purely descriptive ethologist, the natural history phase of the science, drawing to a close. In many situations it is only by experiment that it is possible to work out what is going on within an animal. Lorenz accepts that in principle; but how, in practice, does he react to the more mathematically based studies of ethologists working close to him?

Norbert Bischof was working with ducks and geese at Seewiesen until he and Lorenz both left in 1973. He made studies of Lorenz's greylags and their offspring that involved vast numbers of detailed measurements for analysis by computer. In this study Bischof adopted (and adapted) the method of systems analysis used in control engineering, tracing a network of causes and effects between the various individuals. It was a late attempt to restore the original, intended balance between Lorenz's observation of what is happening in animal behaviour and its detailed causal analyses in the style of von Holst — with, inevitably, the sort of hidden tensions which that would involve. Bischof's experiment with Lorenz's geese successfully tested a computer 'model' of the breakdown of attachment and fear as a swing in the balance of 'forces' pushing and pulling a gosling between family and strangers. He was concerned with the dynamics of the setting up of the incest barrier, the mechanism which in most vertebrates prevents mating between brother and sister or parent and child: on reaching puberty the goslings begin to move away from their own family and become more involved with strangers — processes of social interaction which are clearly of great interest to Lorenz. Bischof had started ('of course') with general observation of growing goslings within their family groups. Like Lorenz, he saw the goslings distinguishing between alien members of their own species and familiar individuals upon which they were imprinted. Initially, the aliens arouse fear while the familiar animals inspire friendliness or attachment, but at puberty, that behaviour reverses: goslings begin to behave as though they have had enough of their family, while strangers become objects of fascination and possible mates.

Such observation produces both the question and a first guess at its answer. But it does not provide any verification of that answer and this is where the measurement comes in, for the change in attitude at puberty is reflected by a corresponding change in the distances and angles at which the goslings sit in relation to the position of parents and strangers. This was observed and marked out on copies of a ground plan over a long period during which two families occupied the same pen, so that a thick pile of observed layouts was accumulated. Then the positions and angles of the geese and their goslings shown on each plan were fed into a computer that analysed the developing situation, removed irrelevant variables and extracted the desired information about the speed of erection of the incest barrier and the spread of ages at which this occurred.

It is significant that Lorenz raises no objection to this work; he respects it because he is satisfied that Bischof is also a good observational scientist who knows his geese thoroughly. Measurement allows the exact study of detail; at any given point in a study precision and comprehensiveness are alternatives, but both are necessary if the study is to be complete. In Bischof's study the dialogue between man and machine is a valuable component of his experimental method: the machine assimilates the drudgery of the scientist's job and lets him get on with the interesting part. Bischof and Lorenz, however, agree that far too many studies actually start with the question 'what can we feed into the computer?', an unscientific question which imposes intangible limits upon the work that can be done, and may lead to a wrong starting point.

An exact contemporary of Lorenz and also from the city of Vienna is the philosopher Karl Popper: the two, says Lorenz, were boyhood friends. Working in England, Popper has had a strong impact on British science through his influence on many leading scientists. An important proposition in Popper's philosophy of science is that each scientist has a prime responsibility to try to prove himself wrong. A hundred different experiments designed to support a theory do not prove it to be true. A hundred different experiments which attempt by every means to disprove it, and fail, make the theory strong.

Quite apart from the philosophical justification, there is good common sense in this. It is a waste of time and reputation to launch into the world ideas that may be readily disproved; and it is an imposition on others to burden them with proof or disproof; besides which, not everyone is a Lorenz, to attract followers who are prepared to spend time investigating not their own but another's ideas. And yet there are many scientists (including Lorenz) who, in moments of true self-criticism, would admit that following Popper's prescription is painful. We are not so made that we wish to prove our theories false; we would much rather challenge an opponent's disproof. Lorenz claims that despite this (and being aware of it) he does try to test his ideas. No idea should be assumed to contain more than a provisional truth, and only by probing it can the weaknesses be discovered, thereby paving a way towards some better approximation of truth or, at worst, clearing the ground for new ideas. In Popper's scheme of things, the source of a hypothesis is not particularly important; indeed, the more extraordinary an idea, the greater its value if it survives disproof. But this would seem to discount any premium on skill: picking a horse with a pin requires no knowledge of racing form and is a poor way to predict a winner. But if that horse wins, the odds may be better than if skill had been used. In real life, however, scientists do not select theories at random; even the most outrageous hypothesis depends on some insight.

Lorenz departs most emphatically from Popper on the question of the source of scientific ideas, or rather, he adds a front end to the circus horse. Lorenz's end feeds on what he considers to be the true food of scientific advance; Popper's end digests or eliminates. Lorenz's method is to observe and then to sleep on it; the human brain has a vast capacity for the storage of data, though only after an initial filtering process. In time, our processes of perception pull the pattern out of apparent disorder, our minds make the connection, and the schooled mind of the scientist eliminates vast numbers of potential patterns without effort. In the introduction to his second volume of collected papers, Lorenz notes that Popper, his authority in other matters, nowhere mentions this vitally important process in his works. Nor, says Lorenz, does he quite understand Popper's rejection of inductive processes as a means of cognition.

Lorenz's papers are long and discursive; in fact, they are barely recognisable as scientific papers when compared with the usual content of learned journals. This is partly because many journals demand the reduction of description to the minimum required to understand the reported experiments (which not only militates against Lorenz but also prevents the reporting of potentially valuable observations, the significance of which is not yet understood). But even more, it is because for brevity and clarity (as well as to conform to the currently accepted intellectual style) most papers today are written as though the author had deduced his results by a logical process that began on the firm foundations of itemised, existing knowledge, upon which were constructed his own careful observations and experiments, building stage by stage until the edifice was crowned by the final result. But in real life (and here Lorenz and Popper are again one), the reverse is often what actually happens: an inductive method has been used, the result has been thought of first and hung precariously in space until a supporting structure can be built to reach it. If this is achieved and the writer of a paper can satisfy a journal's referees that he has taken reasonable measures to knock it down and failed, his paper is ready for publication. What is concealed as though it were scientifically impure is that there was a process of inspired guesswork.

Following ideas introduced to him by his teacher, Karl Bühler, in the early thirties, Lorenz calls his method 'Gestalt perception'. In rough translation, Gestalt means shape, aspect, or form, but add 'perception' and there is even less a simple English expression for the concept. Perhaps the nearest is 'intuition', but this lacks the philosophical dignity of the original, so the German word has been accepted into English.

Gestalt psychologists propose that the brain perceives a unity of form, a pattern it extracts from disorder: the gestalt. At a time when it seemed impracticable to probe the physiology of mental processes and the study of their development had not yet begun, the gestalt psychologists displayed great ingenuity in formulating rules for the setting up of a gestalt: there were as many as a hundred in the heyday of the game. Of those that have survived to be tested in developing infants, one is 'the rule of common fate' which states that moving contours are seen as external edges of a single moving object. Another, 'the rule of good continuation', states that anything that can be described by the same equation in a co-ordinate system will be seen as the contour of a single object. Studies of babies now indicate that these two capacities may be innate, but a third rule—that when several contours are present those that are closer will be seen as contours of a single object—seems to appear at the age of one year. Gestalt psychology was born in Europe and exported to the United States where for a while its complexities attracted many American psychologists. Lorenz has been less concerned with all the complicated theorising, but concentrates on it as a phenomenon, simply describing what it does.

Once seen, a gestalt maintains its integrity in changing conditions as clearly as does a solid object held at different angles or seen under coloured lights. The apparent changes are discounted without thought; indeed, they strengthen the gestalt for they test it by extending the range of conditions in which the perceived object remains constant. It may even continue to be recognised if it changes its form within acceptable limits. The creature perceived as a goose will remain a goose through a vast range of different behaviour. If it suddenly vanishes, the brain will seek for evidence that it has dived below water or flown behind a tree; but if it turns green with yellow spots, says goodbye and then explodes, the observer will doubt his senses or suspect a practical joke: certainly he will feel disoriented.

The constancy (as understood by Lorenz) may take the form of membership of a group or class: a child who has seen a St Bernard and a foxhound will recognise a chihuahua as a dog; and a chihuahua will recognise a young St Bernard as a puppy despite the great difference in size because the creature it sees conforms in shape and behaviour to a puppy: the dog perceives the puppy 'gestalt', which in this case releases the appropriate protective response in the dog. In such examples, behaviour is an integral part of the gestalt, as it is with other forms of human understanding and though it may be seen as more acceptable in artists than in scientists, it is equally vital to both. Lorenz asserts 'seriously and emphatically that in our first tentative approach to the understanding of complicated living systems, the "visionary" approach of the poet—which consists simply in letting gestalt perception rule supreme—gets us much farther than any pseudoscientific measuring of arbitrarily chosen parameters'.

When explaining the process, Lorenz takes the car engine as an example to show the way we understand how a system works. 'The piston sucks a mixture out of the carburettor' . . . and already you have introduced two new ideas that can only be described in terms of other unknowns. The piston as a part can only be understood when you know it is linked by the connecting rod to the crankshaft, and as the crankshaft turns it lowers the piston to create a suction. You can only understand that this draws a gas in and holds it there if you understand the function of the valves, and that the inlet valve is open while the piston draws back; and you can only understand the action of the valves when you grasp how they are operated by a camshaft rotating at half the speed of the crankshaft to lift the inlet valve at the right moment . . . and so on. In fact you can only properly understand the action of each of the parts at the moment the entire system is understood. The human brain can accept and retain information on the elements, reserving full understanding of the role of each part until it finally, in one giant step forward, 'sees' the whole. (But this is not to say that the sub-systems — like the carburettor or the mechanism for firing the mixture — cannot be at least partly understood before the whole structure is seen in relation to its function: in extremely complex systems this may be the only way to approach an overall picture.)

As usual, Lorenz's explanations of the process are liberally adorned with his own graphic illustrations. I once watched him introducing his ideas to a group of students. To their amusement, he rapidly sketched a caricature profile of himself on the blackboard, 'not too similar, but you can recognise it'. He then rubbed out the generous, convex nose and replaced it with a concave, pimple-tipped object. Immediately the similarity that was clear in bushy eyebrows and beard disappeared: it was not merely the nose but the whole portrait that was no longer recognisable, for it is not the collection of parts but their interaction that is characteristic. Similarly, the failure of a single part of a system may destroy its whole function.

The point that Lorenz was leading up to is that gestalt perception is a valuable capacity that can be trained for positive use, rather than totally denied as is currently fashionable in science. Even so, those who deny it are often using the method whether they accept it or not, because this is how the human mind works. Reflecting the views of the theoretical physicist Max Planck, Lorenz comments that this is very like the way a child learns about his environment, and our understanding of the physical world is similarly based on our human perception. As biology rests on chemistry which in turn rests on physics, so physics itself is revealed through our own processes of perception to become what we call knowledge, and this in turn rests on the biology of the brain — and so we come full circle. The simple, practical point is that gestalt perception works and should be a recognised tool in the process of assembling scientific knowledge.

Observing himself, Lorenz believes that his interest in a set of phenomena grows as a result of the unnoticed workings of his powers of gestalt perception, and that what he sees begins to fascinate him when the pattern is already beginning to emerge. As a result, and perhaps without conscious effort, he concentrates more and more attention on the various factors in the relationship until finally the information flowing into his senses has so strengthened the pattern that it emerges suddenly—with a cry of 'Aha!' (Lorenz is indebted to Karl Bühler for his description of the 'Aha!' experience.)

Lorenz rarely attempts to disprove his own ideas by experiments of the type that would satisfy those who honour Popper. Although lacking this formal proof by the failure of disproof, he is happy enough for his followers or co-workers to do it for him. Lorenz at work demonstrates a capacity for gestalt-spotting ('gestaltsehen') that is the envy of his colleagues. It carries him far beyond the level of hunch-production and seems to incorporate the first attempts at disproof essential to scientific method. As the pattern takes shape, any inconsistency is not disregarded but actively cultivated; beside the regular beauty of the gestalt, an inharmonious element jars and jangles insistently until a place is found for it in a modified or new gestalt. Either that or it must be accepted as a severe limit to the 'lawfulness' that Lorenz believes he has observed.

Lorenz feels himself under pressure from scientists like the behaviourists, and even some ethologists who describe him as 'just a naturalist'. But 'naturalist' need not be taken as a term of contempt. In every science there are great naturalists, men whose experiments work more surely than those of their fellows, men whose calculations carry them with remarkable ease to apparently unexpected conclusions and who seem to have a gift for asking the right question at the right time, or for simply being in the right place when something happens. They may achieve their results with deceptive ease, though not always quickly. The great naturalists take a real and obvious pleasure in their work; quite often they are largely self-taught. Such men have much to offer their sciences and their common quality is a sort of scientific intuition – Lorenz's gestalt perception, in fact.

In view of the fact that really good gestalt spotting seems a relatively rare talent in science, is Lorenz right to place so much emphasis on it? Certainly his plea to regard it as a scientifically respectable source of ideas is reasonable, and it might persuade more scientists to put it to more productive use; among medical practitioners, indeed, it is a highly desirable quality. But only an extreme Marxist or behaviourist would say that everyone has the same potential for this, so Lorenz cannot reasonably expect many people to be as good at 'gestalt perceiving' as he is himself; indeed, he is well aware of the problems of communication between those who rely on it heavily and those who do not. It is one of his own greatest qualities, well developed and well exercised, no doubt, but based on a strong foundation within Lorenz as an individual. It could be that a greater formal acceptance by scientists of the value of this sort of mind would attract to science a higher proportion of those more imaginative people who currently regard themselves as suited only to the arts or humanities, where a greater openness to unusual ideas is supposedly more appreciated.

Lorenz loses few opportunities to put his point of view before an academic audience. One anecdote told by Sir Peter Scott recalls the time when, as Chancellor of Birmingham University, he was able to bring Lorenz to England in 1974 for an honorary degree. For such proceedings a timetable is carefully worked out in advance, but schedules can slip despite the best of efforts. And so it was that, as Sir Peter finished his own address, five or ten minutes had already been lost. Lorenz, magnificent in his red-hooded robe and floppy hat, rose to reply. He took his notes from an inside pocket and placed them, folded, on the table before him. He took out his glasses but did not put them on; instead, he began to talk of the splendid new Chancellor who like himself had the gift of gestalt perception—so that he was soon well launched on this favourite subject. Finding the glasses in his hand, he displayed them before him this way and that as an illustration of what he meant: however he held them (he said) the audience could not fail to see what they were. The Vice Chancellor, concerned with the passage of time, could indeed see what they were: to him they were a pair of glasses in Lorenz's hand and not before his eyes, with a set of notes unopened and therefore still to be read, and a programme now fifteen minutes behind schedule. Couldn't Peter Scott do something about it?

In the event nothing was necessary. Lorenz paused in mid-flight, added, 'Well, that's what I wanted to say', put the still-folded notes back in his pocket and sat.

Gestalt perception has snags as well as virtues. Even the combination of previous knowledge and observation can lead to a wrong conclusion. But the naive eye can be in even greater danger, for the mechanism is so powerful it can find and then persuade the observer to believe in order where there is none. If we gaze at the sky long enough or have them pointed out to us, we see patterns in the arrangement of the stars that we call constellations. These may originally have been identified to help the ancient navigators. We now know that (apart from the ribbon of the Milky Way) these patterns are subjective projections on a virtually random distribution, but once we have seen them they cannot be unseen. In time, the practical need of the navigator has been transformed to provide the hopeful correlations of the astrologer; and the

o

astrologer in turn is not contradicted because it is so easy to extract an order from events to match forecast or oracle.

Lorenz recalls a psychological experiment initiated by Alex Bavelas where subjects were asked if they could discover cause and effect between buttons they could press and signals that, unknown to them, were actually presented in a totally random manner. There was a light that flashed or not, ostensibly to tell them they were right or wrong; that, too, was randomised. Nine out of ten of the tested subjects believed they had seen some significant correlation, and one man was still arguing long after the nature of the experiment had been revealed to him that the randomiser must itself have been at fault, because (he claimed) it was unintentionally introducing a consistent pattern of its own. The power of the ordering process can apparently press data into places where it should not be in order to achieve a better 'fit'. If two alternative hypotheses are possible, the information that falls between them may be pushed in either direction to conform to one or other solution, depending on which was 'seen' first. As a result, the first idea that crystallises in the observer's mind may achieve a spurious strength when compared with that which comes later—perhaps too late, because the second possibility has already been robbed of data that might have supported it.

There can be no doubt about our desire to believe in any gestalt we can draw from the world about us. We perceive and believe in the objective existence not only of solid objects, but of strange philosophies that depend upon the strength of an underlying pattern, and we are ready to reject all that contradicts that order, once it is established. So even today there may grow all kinds of new mystical and religious systems that have rigorous internal consistency, but which are inconsistent with each other. And, as Tinbergen has pointed out, the intuition of a poet or novelist may lead him to select a wrong interpretation—'so that the scientist comes in as a verifier and a refuter'.

Of particular irritation to scientists are the pseudo-scientific theories for which exhaustive and exciting evidence can be produced at bestselling book length to the awe of vast numbers who, lacking contrary evidence, can easily be persuaded to see the same order in cosmic events as the person to whom this pattern was revealed (quite possibly by the power of his own gestalt perception). The difference between pseudo- and real science lies in the latter's much broader base. The greater knowledge and expertise of the trained scientist furnishes him with a great quantity of data that cannot be reconciled to the grossly false gestalt of the pseudo-scientist. The general public, of course, has no means of distinguishing between pseudo- and true, and is therefore inclined to accept them as equally valid alternatives, until they are tipped towards rationality by scientists who are prepared to explain what they are doing — or away from it by the systematic order offered by some alternative. In many bookshops that would regard themselves as respectable, 'science' and 'the occult' are offered equal space, and may even face each other on the shelves. In the circumstances, it is hardly surprising that many scientists are extremely cautious of this human capacity for pattern recognition, whatever it may be called and however it may manifest itself; nor is it surprising that many scientists deliberately put blinkers on their own senses.

Lorenz answers that since gestalt perception is a vital component of cognition, the better course would be for each researcher to take this into careful consideration, to be aware of his own capacities in this respect, and to use his known strengths and avoid his known weaknesses. The more complex the gestalt, the more vulnerable it is to distortion, but this can be compensated by feeding into it more and more information – some, preferably, from a different viewpoint. Optical illusions, particularly those where alternative geometrics flip back and forth with the brain itself, are failures of gestalt perception to give the correct answer. It is also reasonable to require that results be tested by traditional methods. Such a finely tuned capacity can be triggered by very small and possibly false stimuli, so it is important to allow that a perceived gestalt can be dismissed by rational argument (although Lorenz recognises that this may be difficult for the research worker himself to do). It is desirable that scientists with differing capacities should approach the same truth by different paths. It is entirely legitimate that where there is a vast amount of complicated data it should be treated in such a way as to simplify it. Lorenz has gone on record as saying that the first glimmer of a gestalt may appear in a graph, a common form of pretreatment of data. Just as much, it may appear in computer output.

At another level Lorenz himself can be attacked. Who (apart from his followers and others who may more rigorously investigate his statements) is to say whether his perceived picture of man's condition is genuinely a first approximation to the truth, or totally false? He is disarmingly enthusiastic in conversation: the questions, he says, are complicated, and there is no proof as yet — but he expects to be proved right in time. The sum of Lorenz's perceptions of the world of animal and man produces a different world-picture from those of his opponents and some of his fellow ethologists. In the process of generating components of that picture (where each new element may be influenced by others close to it) followed by reselection from the result as he writes the books of his later years, is it possible that he has created a structure which is far from reality? His opponents accuse him of this, and by the nature of his method there is little beyond philosophical argument—and more re-assertion—that he himself can offer to counter the attacks.

But his friends can both defend and praise him. Tinbergen wrote to me:

> It is often said that in theorising he goes way beyond the facts and does not bother to record objectively, to quantify, to measure, to experiment. Of course all that is necessary, but where would the be-havioural sciences be if Lorenz had, so to speak, wasted his particular gifts as a pacemaker and a visionary on verification (or refutation) of every single idea? The inspiration of much painstaking research done even now, and in so many parts of the world, can often be traced back to bold—at the time and even now—controversial statements by Lorenz. It is true that Lorenz does not explicitly measure and experi-ment, and he even has an exaggerated respect for my own brand of experimental work; but of course his generalisations do depend upon unconscious counting ('whenever a gander selects a mate, he becomes aggressive to other ganders') and he uses the naturally given changes in circumstances as 'natural experiments'—extracting the general rules underlying the animal's variable behaviour.

And of the need for good observers like Lorenz in ethology, Tinbergen has written, 'We should attract such men, for they are rare'.

Chapter 14

The Homecoming

The year 1973 was yet another turning point for Lorenz. It was, of course, the year of his Nobel Prize, but it was also a year for the breaking of bonds and renewing of purpose. Because he felt there was much more work in him he had no wish to retire from Seewiesen, but that he should leave during the year was decided for him.

The coming blow was first softened by the anticipation of return to his loved home country, an honoured and favoured son at last, and the sting was finally removed when he was presented with the opportunity for new work in desirable conditions, some at his home in Altenberg and the remainder in a delightful Alpine valley about half-way between Munich and Vienna. It took both push and pull on Lorenz to make this transition work, the pull engineered by his old disciple, Otto Koenig, who, for a quarter of a century, had faithfully kept scientific house in Vienna against the day of the master's homecoming and who now made preparations for his return—while in Germany the Max Planck Gesellschaft fêted him during what they had decided must be his final year at the Institute in Bavaria.

At the beginning of 1973, in Lorenz's seventieth year, the eventual success of these changes was a matter for conjecture, but ethology itself was now secure, an accepted and growing science. The ripples of Lorenz's influence had extended far: the study of animal behaviour according to other precepts still exists, but is already changing under the pressure of ethological ideas; and outside the more cautious world of science and its practitioners, the general nature of these same concepts has been widely and readily absorbed by a vast lay public. This in turn has created a sympathetic environment for animal lovers whose work is in the Lorenzian mould, although such a generic description probably does less than justice to many men and women who, like Lorenz, are often characterised by a strong individuality.

One example lies in the disparate set of individuals who have devoted themselves to the preservation of wild species in an attempt to reduce the impact of man on this aspect of his environment. Nature itself seems prodigal enough in both the generation and extinction of species; before man came on the scene the natural turnover was already rapid, and some might argue that the effect of our helping hands in the termination of species in general has been marginal. Nevertheless, the conservation of undomesticated species (and particularly of the mammals which are not easily replaced) is desirable, not only for the selfish reason of maintaining a rich diversity in man's own environment, but also to retain the option of restoring lost genes to lines that have degenerated while supposedly in our care.

Attempts at conservation are themselves often fraught with difficulty for a variety of good ethological reasons, particularly where it means breeding endangered species in captivity for eventual reintroduction into the wild; success depends both on understanding animal behaviour and on the nature of evolutionary change. At Peter Scott's Slimbridge, a rescue attempt for the Hawaiian ne-ne goose was organised when the known world population had dropped to forty-two. Two females and a male began the new phase of population growth which was built up with painstaking care, adding in the genes of extra males that were brought in on two occasions over the years. There was anxiety when a down deficiency was discovered, but the genetic determinants of behaviour seemed more firmly rooted. Eventually some two hundred of the substantial Gloucestershire population were returned to the Haleakala Crater on the Pacific island of Maui, where they were successfully interbred with birds taken from the Big Island of Hawaii.

In East Anglia, another naturalist (Philip Wayre) evidently had similar success with breeding otters but encountered trouble with his eagle owls due to an interaction between genes and environment that Lorenz has made clear to all ethologists: in a word, imprinting. The owls imprinted so easily on their human benefactor that his biggest problem was how to disentangle the relationship in order to prepare them for return to the wild.

Such altruistic activities would be impossible without an intuitive insight at least into the foundations of animal behaviour. But even with the best scientific knowledge, the practical ethologist still finds himself bound by cramping limitations on what can be achieved with animals that are held captive, whether by physical or behavioural restraint. That which is most ready to breed under such conditions — even of mild restraint — may not be the ideal progenitor of a line capable of survival in the wild. The still relatively sparse rewards that have been gained from the practical application of ethology emphasise the need for further 'pure' research.

Ethologists may investigate behaviour in four main ways. They may ask about the survival value of specific elements of behaviour (to Tinbergen who compiled the list this ranks as the first question). Or they may want to know how behaviour works, what its mechanisms are (the purely physiological side of this was investigated by von Holst). They may set out to discover how behaviour develops in the growing individual, which is Hinde's special interest; or they may be concerned with the evolution of behaviour—and here we return to Lorenz and his comparative studies. These four questions are not mutually exclusive; indeed, good ethologists may ask all four but can and do differ in the relative emphasis they place on each: this in turn depends on which is their prime area of interest. But setting such debates aside, we must recognise that all may be valuable parts of an ethologist's equipment. They have all been applied, with success, to the study of animals. If we are to move on from this, the next question (another ripple propagated outwards from Lorenz's original impetus) is: can we apply the same four concepts to create an ethology of man that is more than simple extrapolation from animal observations?

One of the first British psychologists to believe we can was John Bowlby of the Tavistock Institute in London. Concerned with the bond between mother and child, he was brought up on the conventional psychoanalyst's wisdom that hunger was a primary 'drive' and that the bond between infant and mother was secondary to this, created by the dependence of child on mother for food. From his own observation, Bowlby was already convinced that the bond was more complex than that, for he had also seen how psychopathic personality development could follow from disruption of the relationship between mother and child at an early age.

By the early nineteen-fifties he was ready for the ideas expressed in *King Solomon's Ring*, and then for Lorenz's paper of 1950 on the comparative method of studying innate behaviour patterns, and Tinbergen's *Study of Instinct*. He found Lorenz and Tinbergen describing in animals just those things that had interested him: notably, the behaviour characteristics which help to create the bond between parents and young. Bowlby came to accept Lorenzian ethology but with modifications: the mother-child bond is not fully explained either by a system of 'drives' or by Lorenz's 'hydraulic model'. Like Lorenz and Bischof, he considered the spacing of parent and offspring. When this is maintained within a certain range, the child (or gosling) is content: indeed, in the presence of mother he will happily explore to a certain distance away from her, but in her absence he moves less: it is as though he does not seem to know the length of his tether unless he has a focal point. Is this the result of opposing pressures? Bowlby thought not: more likely it is governed by a single balanced system, rather like that which maintains a stable temperature in warm-blooded animals (within certain environmental limits). This concept of behaviour, controlled to remain within a certain range of expression, allowed him to avoid the controversial word 'instinct', although the control system, like that for control of body temperature, would be innate. With his goslings at Seewiesen, Bischof reached much the same conclusion.

Bowlby's ideas (as it later transpired) were only a step on the road to an understanding of what appears to be a more complex relationship between the child and his mother or mother substitute, and also with other adults. The too-simple conclusion that some had drawn from his work of twenty years ago — that all young children require the full-time attention of their mothers — can now be modified to allow for the possibility that a mother may also have needs of her own, and that their satisfaction may be good for the child too. Bowlby still has no hesitation in comparing human and animal attachment behaviour, particularly as this is surrounded by some of our strongest emotional responses: love, joy, grief, despair and so on. We may think of these as uniquely human emotions, but Bowlby (with Lorenz) does not: they are evidence of our animal history, adapted to prompt and maintain our own behaviour within appropriate limits.

If we accept this we may also agree with Robert Hinde's suggestion that animal studies can be of use in a variety of ways to those who wish to understand man. First, they provide methods that may be used directly for the study of humans; or alternatively, where this would be unethical, they offer the opportunity for studying similar problems in other species (in this case the results cannot be applied directly to man but may suggest what to look for in human case studies). Next, theories and ideas developed in animal studies may be considered in relation to man. Finally, if an aspect of human behaviour is also seen in animal species then we may reasonably call in question any uniquely 'human' explanation of that behaviour.

Lorenz would add to this that the survival value of an element of human behaviour is easier to evaluate if we can find an animal which exhibits it, and with which the scientist can experiment. He goes further and says that human culture itself can be seen and studied as a living and evolving system, and that it is legitimate to search for biologically based mechanisms which, in their harmonious antagonism of preserving and dismantling structures, achieve the result of keeping human culture adapted to the ever-changing environment. But practical ethology has not yet advanced so far: for most, it seems a big enough step forward to begin observing the human animal as an individual or in small groups.

Tinbergen's own ideas developed to include one special aspect of the ethology of the human child: the method seemed peculiarly well adapted to the study of the form of autistic behaviour called Kanner's syndrome. In this condition a child seems withdrawn from the world except for a limited number of objects in it, lags in speech and other skills, and displays pointless, stereotyped patterns of movement; those who do not recover have often been consigned to mental hospitals. In their joint study, Tinbergen supplied the behavioural theory to interpret the observations made by his wife, Lies, as they compared the autistics with normal children in which similar elements of behaviour appear from time to time. They decided that it was not a genetic fault as some have assumed, but instead, that motivational conflict must lie at its roots, a conflict between an exaggerated fear and the normal desire to explore and make friends. Lacking social reassurance, their fearfulness increases still further, and the children become locked in a blind alley, often refusing even to talk—although, when cured, they may display a rich storehouse of language. Forced into social contact, they throw tantrums which the Tinbergens interpret as the response of a cornered animal; and since this gets the children what they want, their behaviour is reinforced, so they misbehave all the more. In short, the children suffer from failed socialisation, and therapy must be designed accordingly. The Tinbergens recommend a mixture of firm discipline—though not such as to enhance the children's fear—and finding ways to put them at ease. Even so, the process of bringing them back into a shared world is painfully slow.

It is notable that ethologists who turn to human behaviour have often chosen to study babies and children, not so much as partly-developed adults but as highly competent beings who can manage and organise their own worlds with consummate skill. The study of human adults by similar means remains obstinately in the natural history phase of science: fully developed human behaviour appears fantastically rich and diverse compared with any animal, but a start has been made on that too.

By 1973 the various basic approaches pioneered in ethology were being widely applied to animals and, to a lesser extent, to man in many countries and by a range of scientists, some of whom gratefully acknowledged their original inspiration by the concepts and methods of Lorenz, while others, owing as much, were hardly aware of their debt or even appeared to renounce it. Says Tinbergen: 'Many of our young ethologists don't realise it, but they are still working on problems that were first thought out by Konrad. They have forgotten that, and they turn against the old master.' But it is natural enough that many of them chose to question the earlier 'grand generalisations' of both Lorenz and Tinbergen, and in advancing their own studies they have explored ideas which might be debated by the older ethologists—and sometimes have been. This resulting re-examination might perhaps lead to the strengthening of the questioned ideas. Such academic infighting is often healthy for science, and it is reasonable to expect Lorenz to react strongly.

Criticism from a distance will rarely inhibit competent research that is well under way; but at Seewiesen itself up to 1973 the situation was different. The very presence of such a powerful and prestigious personality was bound to be inhibiting to all but those who willingly and closely followed Lorenz's own style, or those who felt able to make a stand for their own line and method of research. It was not that he deliberately sought to exert pressure, merely that a breath from Lorenz was felt as a high wind among those around him, and as a gale by any who were still insecure in their science. Inevitably, it became easier to delay potentially controversial new ideas than to air them while still half-baked, to fragment into departmental sub-groups than to cross the boundaries of the various specialist studies.

Since the deaths of Kramer and von Holst their former departments had been developed along lines that broadened the range of 'behavioural physiology' studied at the Institute. The *Abteilung Schneider* reflected Dietrich Schneider's interest in olfactory insects, and was clearly far from the sort of behavioural study that Lorenz knew and loved: it lacked the warm-blooded involvement between man and animal. (Curiously, however, recent work in America has drawn attention to remarkable and fruitful analogies in the patterns of altruism and aggression shown in the compared societies of ant and man.) The *Abteilung Mittelstädt* dealt with bio-cybernetics and almost by definition set itself apart from Lorenz. This apartness was physical too, since it was housed on a small hill above a corner of the lake, screened behind a wall of trees as though to emphasise that it was inward-looking to the computer terminal rather than outward through the open air to the other departments. Lorenz had originally agreed to the new arrivals, but he must have had second thoughts about the change in overall emphasis that resulted.

As for the *Abteilung Lorenz* itself, the driving force behind its work seemed in time to have diminished as Lorenz had less to do with the day-to-day animal work and more with the writing of his ideas in papers, lectures and books. The days of invigorating debates between Lorenz and von Holst were long gone, and the heady atmosphere of the early intellectual enthusiasms replaced by a sense of withdrawal. By the time the ice on the small lake was thinning and the late snows on the meadow began to melt at the end of the long winter of 1973, Seewiesen was marking time. Individual research continued in the various scattered laboratories—with the vigorous or deliberate pace of the various individual scientists—but the Institute as a whole seemed to hold its breath.

I visited Seewiesen then (in fact, several times between the late winter and early summer), watching the surrounding countryside of fields and woodlands cast off its mantle of white and misty grey for the pastel shades of spring, deepening to richer, maturer greens under the summer sun. The geese and ducks that had waddled uncomfortably on the ice swam confidently in their pairs, competing for the repaired or renewed nest boxes set out in the middle, or laying their eggs in the reeds around the marshy edges. The goslings were soon following their mothers—or Jane, Brigitta or Charley as mother surrogates—around the meadows between the buildings, nibbling at the young grass. Later they sailed out on to the lake itself, sometimes drifting, or turning curiously to explore some new item of interest, then scuttling to restore a safe proximity to their maternal anchor to the world.

Over the years a succession of students or 'goose girls' had come to live with and bring up families of geese, pledging companionship for the goslings from the first 'conversation' with the egg (that dialogue of squeaks and endearments through the yet unbroken shell) to the time when the goslings turned their interest away from their mothers and became self-reliant. Their patient performance of a twenty-four-hours-a-day job that demands unending consideration for their charges gains Lorenz's respect and liking, while he, by his unstuffy attitude, is popular with them. One group made an eight-millimetre film in which, by way of contrast, Lorenz was cast as the absent-minded German academic marching stiffly across the marsh, solemnly reading from a

learned paper held out before him while clad only in a pair of bathing
shorts—very conveniently, since the plot demanded that he should
misplace a step and fall deep in muddy water. The short film concluded
with a few last bubbles rising from the mire, one of which was presumed
to contain the soul of the drowned professor, since as it burst the camera
lifted to follow the invisible vital spark to heaven above. When he saw
this last unexpected detail in the completed film Lorenz roared with
delighted laughter.

His occasional habit of sunbathing with the students and their
goslings at a smaller, supposedly private lake behind Seewiesen led to a
problem with local youths who, it seems, took to ogling such remarkable
proceedings from behind the bushes, which incensed the parents in this
staid country community. Eventually they agreed that recumbent or
seated sunbathing was no threat to the morals of their young, but would
the professor and goose girls please not sunbathe while standing up?

As for the gosling studies, these have been criticised on more orthodox
grounds as producing little of scientific value to match the time and
effort involved. Except in the work of Schutz, critics claimed to see little
forward sweep of understanding that could bear any comparison with
Lorenz's early work—although Lorenz himself was proud of a study by
Jane Packard, a young American who had joined the ranks of the goose
girls seeking to discover whether her own quantitative studies (measuring distances, etc.) and Brigitta Kirchmayr's purely observational notes
would lead them to similar or different conclusions about the behaviour
of a particular family group. But among Seewiesen scientists not directly
concerned with geese, I could detect no general sense of pride in recent
scientific results by which the Institute as a whole might be bathed in
reflected glory. Rather there seemed to be a feeling that it was time for
the geese to go—in part, at least, because they had largely destroyed
their own environment.

At first Seewiesen had seemed ideal for the geese and they had
prospered. It was, of course, quite natural for their numbers to attract
the attention of the local foxes, and these carried off a proportion of the
young. But in time the ready availability of prey led to a corresponding
increase in the activities of the predators until the night Schutz lost
fifty-eight birds to a single vixen and nearly a whole year's work was
destroyed. A double defence line of electrified wiring and fox-proof fence
saved subsequent generations of goslings but accelerated changes in the
lake itself. Now the excess of geese gradually trimmed the reedy margins,
and by loading the sluggish waters with their droppings they over-

fertilised and so destroyed its former balance. In the resulting eutro-
phication, blue-green clouds of microscopic algae flourished and then
rotted, so removing vital oxygen until large pike that once lurked in the
waters now floated still on the surface. It was found that the cost of
restoring the lake would be high — and a waste of time and money if the
geese were to remain. If they were to leave, and their celebrated leader
with them, the lake would no longer serve its original purpose — but then
the Institute itself was also at a point of change.

Of Lorenz's original team a few were still at Seewiesen. Jürgen
Nicolai, with his brood-parasite studies, also Friedrich Schutz; but
Eibl-Eibesfeldt had gone to set up his sub-department of human
ethology a short distance away. Helga Fischer, who with Heidi Buhrow
had managed Lorenz's goose studies for many years, would leave
Seewiesen to follow the geese to their new home. Wolfgang Schleidt had
long since gone to America and remarried, while Margret Schleidt
remained at Seewiesen as Lorenz's assistant. Heinz Prechtl (from the
original Viennese group) was at Groningen in Holland and had reverted
from his temporary ethological interest in lizards to the study of human
babies, pioneering the early detection of brain damage by observation
of behavioural disturbances. He has also shown that breech babies —
those born bottom first — begin their extra-uterine life with different
reflexes from those of babies born the usual way round.

Of those who had joined Lorenz at Buldern and were still with him in
1973, the man with the widest range of interests and ideas was Wolfgang Wickler. Quieter and more disciplined than Lorenz, Wickler might
easily have been missed in the shadows cast by the near-flamboyance
and periodic firework displays of the older man. From Buldern onward,
Wickler had paralleled many of Lorenz's own direct concerns: in his
animal studies in particular, he was interested in the role played by
aggression in animal societies, and the forces — notably pair-bonding —
that balanced it. A perceptive observer of the details of behaviour, he
applied his skills to a broad spectrum of animals from monogamous
shrimps to duelling birds rather than concentrate most of his attention,
as many ethologists do, on a single species or group. He was therefore
well equipped to follow Lorenz's interest in comparative studies of
particular 'organs' of behaviour. But Wickler was very far from being
the slavish follower of his master: he branched out as a thinker to offer
his own interpretations of what he saw, but avoided conflict by specialis-
ing in such matters as the structure of monogamy — very close to the
mainstream of Lorenz's studies but not overlapping them in such a way

207

as to produce a defensive reaction from Lorenz. In fact, the meticulous care of his methods and the intellectual rigour of his conclusions won Lorenz's respect.

Monogamy is a feature of a large number of animal societies. Some have inherited it from common origins, but others clearly have not: monogamy is simply one of a number of elements of social behaviour which can be put together with others to form a stable and viable social structure. A variety of other systems are equally possible and so, in the vast range of natural species, all are seen side by side. Evolution has in time produced a fascinating interplay of the various systems, and in seeking to disentangle some of this rich complexity, Wickler was well equipped to consider the value that could be placed on arguments that depend on analogy (where evolutionary paths have converged to a similar end) compared with homology (where similar behaviour is derived from a common ancestral source from which it has diverged little).

Among the more spectacular species to have been studied at See-wiesen in recent years were Wickler's painted shrimps, their flesh, visible through the translucent shell, brilliantly daubed in pink on white (and for no obvious reason that answered Lorenz's question 'What for?' since the shrimps themselves have no eyes and 'see' their mates or their food by the subtle nuances of flavour these impart to the water around them). Separate the mated pair and their activity is at once increased, as the creatures wave their antennae seeking to ascend the chemical gradients to restore the taste of companionship which alone will alleviate their nervous stress. Crowd them together with a disliked companion, and the increased stress may lead to the death of one or the other.

To determine the roots and nature of the bond between pairs, Wickler had first prepared the ground by establishing that the pairs were genuinely male and female (92 per cent were). He had asked whether the pairing was an accidental by-product of two shrimps showing a preference for the same spot, or whether the preference was for one specific partner. A computer offered the answer that there were preferences for both partner and location, but when forced to decide between the two, a shrimp would choose to stay with the partner. In experiments where partners were mixed up, the shrimps promptly sorted themselves out again. Wickler had carefully elimi-nated the possibility that the bond necessarily depended on primary 'drives' such as fighting or feeding, brood care or sexual preference (the male, it transpired, would readily copulate outside 'marriage'). The answer he found bore striking similarities to the attachment theory that Bowlby held to account for the bond between mother and child.

Wickler's painted shrimps proved difficult to keep in aquaria, so by 1973 after the studies were completed these decorative creatures were no longer seen at Seewiesen. But they were replaced by another curious and (when observed through anthropomorphic eyes) delightful scene: the love duet of Wickler's assistant Gaby Tyroller and the barbet birds imprinted on her. To Gaby's cajoling and caressing call of '*komm-komm-komm-komm-komm*' a bird would reply by flying on to her finger close to her face, dipping its head and calling rhythmically in barbet courtship, tempting the girl with the love-offering of a juicy meal-worm — all oblivious of the fact that the prospective partner was not only of a different species but also of the same sex as itself.

Both bird and crustacean were part of Wickler's comparative study of monogamy in species deliberately selected as being widely separated by evolution, so that the essential common causes in grossly different environments could be sought. In these two cases, the similar behaviour was clearly not homology but analogy, so it was necessary to ask just how close that analogy is: do the two forms of behaviour conform in great detail or are there significant points of difference? Where there are differences in detail, the purpose served may also be different. In the case of bird and crustacean the analogy seems strong, but for each new analogy the question must be asked afresh — in particular, when we compare the monogamy of man and wife in our own society with the monogamy of, for example, gander and goose among Lorenz's greylags. Together with Eibl-Eibesfeldt, Wickler had considered some of the differences between human monogamy and that of various animal societies, and proposed that the analogy is incomplete and therefore could be misleading. Far from monogamy being a human quality which some animals shared, they saw it as a relatively simple and clear-cut evolutionary conclusion in a variety of lower animals, having a survival value in the societies of which it was an integral part — while in man they saw nothing so definite.

Man has a capacity for divergent cultural evolution, and this has led to widely differing societies that are successfully adapted to their environments. Each society maintains the behavioural elements that made for success in survival by enshrining them in religion, folklore or other firmly established determinants of customary behaviour. In the case of human sexual behaviour and marriage customs, widely differing systems have been evolved and maintained under what must be very loose genetic control over the details of the systems chosen by different societies. The only essential ingredient is one that is different from lower animal systems: in man the bond between partners must provide for the care of offspring through a period of growth, development and learning which constitutes an unusually large proportion of the individual's lifetime. Wickler and Eibl-Eibesfeldt concluded that monogamy in man, in those societies which prefer to practise it, is an arbitrary but successful way of achieving this result, and necessary (by any natural law that they could see) only for the period during which the young require protection. They further proposed that continuing sexual behaviour in marriage was designed not simply to continue procreation year by year throughout the fertile period, but rather to help maintain, renew and strengthen the bond between partners for a period long enough for the children to grow up.

The use of the expression 'natural law' in this context is not accidental. The Catholic Church bases its prohibition of contraception on 'natural law'—which should be the same thing as a scientifically observed law of behaviour. The two pupils of Lorenz suggested that a direct application of ethology to man might lead the Church to a revised understanding of the natural roots and purpose of human behaviour that might be more in tune with the needs of an already overpopulated world.

Like Lorenz, Wickler was well aware of the human relevance of his ethological studies but much more cautious about expressing it, except when working in co-operation with Eibl-Eibesfeldt, who had made a comparative study of pair-bonding in many human cultures. Together they could speak of the differences between the essential monogamy of Wickler's animal species and its cultural expression in humans as one part of a wide range of behaviour patterns accepted by different societies.

Pair-bonding, brood care and aggression are all interlinked. If the general laws that seemed to hold for Wickler's monogamous animals could not be applied directly to man, might the same also be true for aggression? On this, while Wickler was a member of the Lorenz department, he could only be non-committal, although it was clear that he suspected that when the comparative method used to study monogamy was turned to the subject of aggression, this too was going to prove different in man. Indeed, even among animals that do fight, much work was still to be done to establish more clearly the true nature of their aggressiveness. Does the inclination to fight really increase spontaneously as time passes since the last encounter? And does the same thing happen even if the animals are reared in a totally peaceful environment? Lorenz suggests that animals with the instinct will eventually fight, although other observations suggest nothing so clear-cut—rather, that some do and some do not. There is one universal inborn instinct of this type: the desire for food which certainly does grow as time passes since the last meal; yet humans can force themselves to starve. . . .

The results of the still necessary studies of aggression in animals and man could be embarrassing if carried out by a department of which Lorenz was still the head. While he had often said it was a healthy exercise for a scientist to discard a false hypothesis each day before breakfast, it is only to be expected that as he gets older the scientist will defend his hypotheses with greater tenacity. So how does Lorenz himself feel about issues like this?

He states firmly that he is well aware of the difference between analogy and homology and in particular of the caution with which analogies must be handled; but equally, analogy remains a powerful tool. Further, he knows perfectly well, for example, that certain forms of family structure, marriage, or incest taboo can be instinctive in animals but culturally determined in man. Ethologists do not blunder often in diagnosing whether a pattern is instinctive or cultural, but where he himself has erred in the past (he says), it has more often been on the side of wrongly assuming that something has been learned when it later turned out to be innate. It is true that he has left a lot of work for others, but as for the hypotheses that remain unproven, the long shots and the daring assertions—some of which even his own pupils refuse to believe—Lorenz is confident that when they get around to working on them they will find that in these matters he was right after all. 'And that', says Lorenz, 'is in a way a very ambitious belief'. It is his simple expectation that like Darwin's his own views will stand the test of time, to be accepted not only by scientists but also by most other intelligent people as simple and obvious truths.

By his seventieth year—his last at Seewiesen—Lorenz, though still active, had inevitably slowed and changed some of his earlier ways. He no longer joined his geese for a swim at six in the morning, although he still followed their example in taking a midday rest and could thus sustain his two activity peaks in the day. These were now used very little for systematic animal observation, although he could still see

P

familiar species with an eye fresh enough to pick out new and significant events. One which surprised him was the way in which the older of his two dogs had taught the younger not to attack geese—simply by markedly not doing it. Such learning from negative example is well known in rats which can follow a tradition of not eating a particular kind of food, but Lorenz had no reason to expect dogs to have the same trick in their repertoire. The older bitch, Claudia, a hunter who could kill a cat with one shake of the head, was obviously tempted by the geese. But having been disciplined not to chase them she now cut herself off from the strong stimulus of their presence by looking away when they were near. The younger bitch, Babette, had evidently learned from this that geese should be avoided. Lorenz said he intended to keep more dogs after retirement from Seewiesen and this was one detail he would follow up. In fact he had six within a year—though not entirely by choice, since the four pups were mongrels. (His sketch of Babette at Altenburg is on p. 109.)

For the time being, however, such observations were little more than an agreeable form of mental relaxation from writing, his main occupation. His study at Seewiesen had a large tropical fish tank set into the wall in one corner and a pair of shama birds nesting on a high shelf among the crowded books. Wearing horn-rimmed glasses for this close work, Lorenz typed at a desk cluttered with papers, more books, and scissors and paste for the frequent revisions of order and interpolation of new ideas into his draft typescript. The subject of his writings was rarely new work on the animals surrounding him; the books, papers or lectures he wrote were nearly always on cognition, the nature of human knowledge, his philosophy of science, or speculations on the condition of man—for none of which did he really need a desk here at Seewiesen rather than at his old home at Altenberg.

During one of my visits he was busy with an address to be given on the occasion of an unusual award, the Paracelsus Ring that he was to receive in the small town of Villach, beyond the main eastward string of the Alps in Southern Carinthia. For Lorenz there was double delight in this honour: firstly, because it represented yet another sign of welcome from his home country, and secondly, because of the name associated with the award. The exploits and achievements of Paracelsus, one of the most extraordinary, tempestuous and controversial characters of the early sixteenth century, might have been taken as a heroic model by Lorenz himself. Theophrastus Bombast von Hohenheim, called the Great Paracelsus by his followers, had sought to sweep away the medieval dead wood of the medicine of his time—symbolically burning some of the overrated classical treatises—and almost incidentally founding the pill-based medical practice of today. He wrote that only man's lack of imagination prevented him from seeing the certainties of the arts and sciences. By the standards of his own time – and in many ways ours too—he was a believer in magic, and also by his wilder assertions paved the way for the name of 'bombast' to enter our language. Yet, if he could see the world today, he might well observe that his apparently magical claim that 'resolute imagination can accomplish all things' was no more than a flowery overstatement of what scientific advance would make commonplace in centuries then to come.

It gave Lorenz great pleasure to be honoured by association with this early pioneer of medical science who had combined natural history with philosophy, and he typed out a lecture to match the occasion. A few days before he said that he proposed to lead off with some ideas on free will. If he considered his own behaviour with scientific objectivity he could not doubt that all the neuro-sensory functions involved proceeded by strict cause-and-effect; and yet if he considered his own behaviour subjectively (subjective experience being the pathway of all his knowledge), he could not doubt his own freedom of choice without denying all that body of subjective experience: '. . . and this is a contradiction with which we must learn to live.' He follows an older philosopher in believing that we do not have, nor should we have, an absolute, arbitrary and unbounded freedom, but that we may seek to achieve a limited but two-fold expression of our will: freedom from external coercion and freedom to follow the ethical moral law within oneself; the latter would, of course, be strongly though loosely bounded by genetic and cultural limitations.

On the day of the award, the address in his honour was ceremonially packaged in a seemingly unending stream of string quartet, greeting, introduction and description of his work to that moment, until in the end Lorenz turned to his wife in dismay and whispered that in lauding him they were covering half the ground of his own address! When allowed to stand up at last, he had to set aside several pre-empted pages of his own prepared speech; that, he said briskly, was an example of free will in action.

His freedom was similarly bounded in the matter of choosing the new leadership of Seewiesen. By his own account, and fearing that the current craze for quantification would stunt the descriptive work still so necessary as a base for science, he fought tooth and nail to bring in someone whose approach was mainly descriptive. Others saw a necessity for regeneration, with the setting of new goals rather than the following of Lorenz's style in a way that would appear like a performance of *Hamlet* without the Prince. Appearances did matter, for the sheer popularity of Lorenz was such that his departure would seem like the end of an era: it must be clearly seen that ethology and Seewiesen remained alive and well. One new direction might lie in the new but more experimental mainstream somewhere between Tinbergen and Hinde; another line that could continue alongside being Wickler's more descriptive, social-ecological approach. For a while, the idea of a triumvirate was canvassed, Wickler being the obvious choice as the descriptive man, to be balanced by an experimental ethologist and a neurophysiologist in the mould of von Holst. In the event, the process proceeded painfully slowly, and the uncertainty surrounding Seewiesen deepened as Lorenz's own future became clearer.

Otto Koenig had been busy behind the scenes: a perfect site was found for the geese in a valley probing southward of Gmunden, deep into the Alps at the Totes Gebirge. Cradled in the head of the valley and protected from extremes of climate was a small lake, fed by springs warm enough to provide some clear water above the flowing stream the whole year round. A pleasant small town half way up the valley, Grünau, had become a minor holiday centre for Austrian and German tourists, thus providing a link with the outside world, while reasonable privacy was ensured by the established use of the lands of the upper valley as a private deer park. A zoo of European animal species (providing a focus for the tourists) would be a valuable neighbour. In short, it would be an ideal location, made even more perfect by the presence of a two-hundred-year-old cottage just big enough for a goose mini-institute, set back on rising ground and looking out over running water—just the prospect that Lorenz prefers.

It was called—oddly, to English ears—the Cumberland Deer Park, a title deriving from the Hanoverian connection. Lorenz went to see the present Duke, H.R.H. Ernst August von Cumberland, tall, good-looking and—here Lorenz found himself joining Wodehouse's Bertie Wooster in a search for the right English adjective: '. . . there must be one ending in *air*'. Finally, he found it: 'Debonair—he's a debonair grand-seigneur, that's what he is.' Lorenz is really no more averse to the company of Royal Highnesses than his father, and was charmed by his host's intelligent conversation, so that both were happy to continue with the arrangements that were now well under way. The chosen house was already being modernised—at the gentle pace of the region—to be ready to receive the exodus from Bavaria at a time when the older geese were moulting but that year's hatchlings were not yet old enough to fly.

On 13 June 1973, Otto Koenig went on Austrian television with the Minister of Science and Konrad Lorenz as his guests in his regular animals-and-environment show. In the rustic log and split-bamboo Wilhelminenberg-inspired studio setting, the plans were announced: Austria would at last be home to one of its most distinguished sons. To the dismay of some viewers, the lady Minister was carrying what looked remarkably like a crocodile handbag, but she nevertheless pronounced the official blessing upon Koenig's welcome to the celebrated naturalist and conservationist.

In administrative fact, Lorenz now would merely be head of Department Number Four (Animal Sociology) within Koenig's Institute for Research in Comparative Behaviour under the Austrian Academy of Sciences. But Koenig saw it differently: to him Lorenz had always been the prophet, and all this had been created for him. It was like a church whose god came down to live within its earthly structure, not to displace the priest from his former role, but hardly to be considered a subordinate.

To transport his geese to their new home took several journeys, and late in June I joined Lorenz for the move of a batch of geese together with many of that year's goslings. We assembled at six on a bright, sparkling morning. Lorenz, loath to leave, busied himself briskly with the mechanics of departure. Then it was discovered that one flock, penned overnight ready for the move, had unthinkingly been released and was now irretrievably scattered over the lake. There was to be a big press conference the next day in Austria—and all would be there but the geese! After a rapid change of plan, the emotional build up of the leave-taking together with dismay at this new development was abruptly discharged in the vigorous hilarity of a literal wild goose chase. After a good deal of running, encircling and pouncing, some of the pairs (made less mobile by the need to stay with their own small families) were rounded up and loaded on to the two minivans with the boxes of imprinted goslings which would travel in the charge of their 'mothers'.

In the now lightened atmosphere Lorenz turned to the men, women and children who would remain at Seewiesen; he spoke to several, kissed a small girl, shook hands with Wickler, and as the almost feudal line waved him on his way, the two minibusloads of geese and their attendant cavalcade of assistants and well-wishers sped out into fresh, early-morning countryside and through the clean, cheerfully frescoed Bavarian villages to the autobahn and the frontier at Salzburg. Here there could have been problems, for a bird that could quite easily fly unchallenged over any frontier would quickly find itself in quarantine if it had the temerity to walk (or be carried) across. But Lorenz's famous geese were not to be fussed by regulations. Koenig had sent a vet with papers that conferred upon them a diplomatic immunity from disease. The customs officials read them sagely, peered into the boxes, stepped back and saluted — and the geese had become officially accepted residents of Austria.

In the valley of the Alm, the minibuses turned from the paved road to a track through the grounds of a sawmill that would serve as a defensive outwork against unwanted visitors, bumped over a rustic bridge and up to the door of the future home of the Institute, on which work was inevitably still progressing. The goslings were unloaded to a chorus of soothing and reassuring noises from their 'mothers' and led away to inspect meadow and pond, while Lorenz helped to decant the geese into a large pen that was half in the stream: here they would become accustomed to the unfamiliar surroundings. As Lorenz up-ended a box to extract a final reluctant bird it suddenly fell awkwardly and was hurt. With distress etched into his face and eyes, he swept it up and away for a first-aid inspection. Later, with the bird attended to and Lorenz satisfied that all was well, he laid himself down on the rough, sloping meadow below the trees for his afternoon nap.

Konrad Lorenz had come home.

Chapter 15

The Prophet at Home

Back in Austria, Lorenz recalled a favourite Kipling poem about prophets having honour all over the the earth except in the place they were born: *But O, 'tis won'erful good for the Prophet!* Now, it was all the more a joy to him that at seventy he was truly recognised in his own country and could at last return to it to work.

At the goose mini-institute there was no shortage of work for his assistants. After the first two summers, the scientific 'protocol' recorded that, of two hundred and forty-eight birds brought from Seewiesen or hatched in the following year, forty were dead, sixty-two had flown away and not been heard of since, and twenty-three had gone though their whereabouts were known. Some which had made several attempts to get back to Seewiesen seemed to decide at last that their home waters were jinxed and settled for another lake a few miles from it. Flight times varied from two days upwards: one bird clocked in at three months. Some left the Alm valley on exactly the same date two years running, arriving at Seewiesen by the same schedule. The snow geese tried several lakes around Seewiesen, and for three summer weeks feasted daily in a field of maize. Brigitta Kirchmayr recognised some of her own 'children', who had missed by a greater margin, in Munich's public park, the *Englischer Garten*.

They flew away for all sorts of reasons, some simply by mistake, when they were disturbed and became frightened in the night or did not like the look of the snow. At Seewiesen, a bird flying away was nothing to cause a moment's concern for everyone knew it would return. But here the surroundings were strange: not only were the geese far from their accustomed home, but they saw about them terrain that geese would not normally choose, with mountains hemming them in. To watch a bird flying away here was dispiriting: you might never see it again. So pleasure was all the greater when, after prolonged absences, some of the dropouts dropped back in again to rejoin the flock in ones or twos.

Although the old trouble with foxes looked like repeating itself, their new environment was in most other ways healthier for the birds, and the springs in the Almsee kept their promise, leaving a sheet of water as big as the old Ess See clear through the whole winter. The geese that had learned to fly in the valley were happy enough to make it their home, but of the oldest birds from the years at Buldern there were few remaining, just Moritz (hatched 1952), Adonis (1954) and Sinus (1955). The shifting Institute had outlived or shaken off all the rest of its oldest inhabitants long before their time. And yet some useful observations were possible: it could be seen that certain patterns of behaviour changed very quickly. Birds that would not eat from the hand at Seewiesen became more dependent and would do so at Grünau, and watchers could note how fast new habits of flight were set up to avoid areas where predators lurked; and the goose mothers could continue their own separate studies with their growing broods.

At a press conference at the start, Lorenz had disclosed his plan to include wild pigs and beavers along with the geese in new comparative studies of animal societies, but in the event money was tight and these were slow starting. A year and more later, there remained only a nucleus of the goose-keepers to hold the fort through the winter, plus one young man watching over the activities of two pairs of beavers. Because the Alm valley was strictly limited to European species, only a part of Lorenz's renewed animal research could be accommodated. The rest, his tropical fish and even the pair of shama birds from his study at Seewiesen, would have to go to Altenberg to join his dogs Claudia and Babette and the Danube fish as the nucleus of a newly animal-populated home where Lorenz would now be for much of his time. There, an important feature would be a new aquarium building at a far corner of the garden. It would have less of the small tanks that at Seewiesen had proved too constricting for territorial fish to be properly studied; instead, the main tank would be the size of a room in itself, with territory enough for several pairs of butterfly fish. This, at least, was the aim: in practice it has been a long time getting under way, for when all was settled the contractor was called away to higher things, the building of a church.

This new fractional Institute would come to an Altenberg changed and still changing from that of thirty and forty years before. The twisting, constricted road through the villages remained to daunt travellers who might prefer broad featureless freeways, but plans for a future autobahn threatened even here on the bow of the river. The

218

nearest point in the water meadows, where young Konrad and Gretl had played ducks and drakes, was now a dumping ground for earth and rubble, while the scrub beyond had become crowded with ugly holiday homes on stilts; otherwise, post-war prosperity seemed to have by-passed Altenberg.

Frau Dr Lorenz meanwhile had redecorated and renovated the house for its master's second return. Over the months, the faded rooms brightened under new paint, and Gretl rummaged among the pictures for those her taste would allow to remain. Did not I too think they should go, to uncrowd the walls, or be replaced by pieces of greater individual merit? Of that I could not be sure. The specially painted canvases at least offered a unity of style with the house and were enjoyable for that, and Konrad himself evidently had no objection to them. But out they went, one immensely long picture pausing momentarily in its outward progress to screen Adolf's 'crazy' hall from an intrusive slash of sunlight while I filmed that which remained. Apart from the *Diana* of hopeful provenance who continued to reign darkly to the left of the fireplace, one canvas that could hardly be removed was the ceiling adornment, *Victory of Peace Over War*—that celebratory climax of Adolf's architectural folly. Also remaining were *The Four Ages of Man*, set into the wall above the staircase complete with the naughty weeping child Konrad. In the end, Gretl's rearrangements worked; the hall was improved while sufficient of Adolf's cheerful vulgarity remained to keep its atmosphere intact.

The temporary sterilisation of the house during its refurbishing was emphasised by the silence under the eaves. With Lorenz I climbed the attic stairs and along the corridor of twists and turns made necessary by the bulk of the central hall, to the tiny room where the young Konrad had kept his first birds. We went through a door to the still, clear space under the rafters that had echoed until so recently to the raucous jackdaw calls. One small gap in the under-roof planking marked the site of the first nest. Two forlorn boxes still hung by an open window, and a screen of wire-netting that had long ago been tacked up to retain a region for the attic's traditional role as junk depository now separated discarded furniture from emptiness. Lorenz looked about sadly at the unaccustomed free area of floor boards, his eyes rebuilding the piles of twigs and droppings that had accumulated over the forty years from 1928 to 1968 but were now gone—due, in his view, to the use of agricultural insecticides.

That morning, in search of the one remaining colony in the district, we took Lorenz's Mercedes over the Danube by the old Klosterneuburg vehicle-ferry that is strung from a cable stretched over the river, and claws itself briskly across by the force of the swiftly flowing waters against the angled prow. From here we drove up to a romantic, reconstructed mock-mediaeval castle on a lonely hill. On the battlements and turrets of Burg Kreuzenstein, the memory of the lost colony came alive to Lorenz under the swirling and eddying streams of jackdaws, sometimes visible only as an ever-changing pattern of black particles high above us, sometimes sweeping down to settle in ones and twos across the ridges of roof or gable, then darting away again in unison, and all for no evident reason. The air was alive with their mixed chorus of 'ja' and 'jaw', the calls for going out and coming back, that were raised in conflicting antiphony. 'That's what you call democracy', commented Lorenz.

In the grass of the courtyard lay a dead jackdaw on which he found evidence of poisoning. Then came distraction by a quieter, shriller sound and he pointed to the tiny slit of a lancet window in the tower high above the courtyard. 'See there! A bird coming out of that window. There must be a nest in there'. By a series of flights and spirals we climbed, and sure enough the small hollow of a nest seemed cemented there in a vast pile of twigs which lay inside the open slit. None too expertly Lorenz mimicked the call of the parent, 'tchock — tchock — tchock', but it was enough to renew the squeaky chorus and raise up gaping mouths. As he picked up the hatchling and allowed it to bite on his little finger, more than forty years fell away. . . . 'I'm getting nostalgic about jackdaws; maybe I'll settle a new colony at Altenberg.' When he repeated this later to his wife, Gretl did not hasten to encourage a whim that would probably get lost in the rush and fuss of all the other activities surrounding his homecoming. The rotten timbers had only recently been replaced.

New jackdaw colony or no, Lorenz was now re-immersed in his original world of animal studies — or, at the very least, the preparations for it; his past experience as a creator of study-groups and institutes would stand him in good stead, though it also meant his becoming a commuter, as Altenberg and Grünau are separated by a good half-day's drive and a car is the only practicable means of rapid travel between the two. During the next year, his seventy-first, he could set himself a far busier programme than before his so-called retirement. Added pressure, mostly but not entirely welcome (because it reinforced and focussed the fire of his detractors and political critics), came with the news that on 10 December the new Swedish King Carl Gustav XVI would honour him with a shared Nobel Prize for Medicine. One of the unwritten Nobel rules might be 'live long': this qualification was certainly obeyed by his fellow laureate Karl von Frisch, just seventeen years Lorenz's senior, and no longer able to make the journey to Stockholm. The patient and conscientious deliberations of the Karolinska judges had brought them to the conclusion that his life study of bee communications made von Frisch the founder of a field of modern ethology which was complementary to that of Lorenz and Tinbergen.

A firmly written rule of the prize, deriving from Alfred Nobel's blinkered view of scientific values, restricts the award to a narrow range of sciences and to specific discoveries or inventions within these prescribed bounds. While even mild critics of Lorenz could applaud the award of this accolade, many professed astonishment that he should have been considered as qualified under the rules. The judges, to their credit, had interpreted the out-dated restrictions liberally. The argument in his favour may have proceeded as follows: first, the prize is that for physiology as well as medicine; second, animal behaviour (according to Lorenz) is a physiological function with its own unseen but nonetheless real organs; and third, what is learned from the behavioural physiology of animals may have important applications to man—and so, potentially, to human medicine. The award was as much a triumphant vindication by the Nobel judges of Lorenz's own definition of the nature and relevance of his science as it was reward for a set of specific qualifying discoveries that were by now matters of scientific history.

Twice before, the supposed secret that his name was before the Nobel judges had seeped back to him, and on the second occasion, some two years previously, journalists had even come prematurely to congratulate and interview him. But in 1973 he was 'third time lucky'—and it did come as a surprise. He was sick in bed with a sinus complaint that would trouble him for the next year, and Gretl had gone to bed too, when the telephone rang. It was a journalist asking for an interview. Gretl told him firmly: No, her husband was ill in bed. Perhaps, the journalist suggested, there might be circumstances in which Lorenz might be prepared to come to the telephone? Gretl repeated: No, she was quite sure there would not. The man persisted; didn't a call from Stockholm suggest something? Then the light dawned. Maybe, Gretl conceded, Konrad would give a telephone interview. Soon afterwards, a neighbour who had heard the announcement on the radio rang to tell them the news was public. German papers reported that Lorenz perked up enough to remark 'not without humour' that his father would have been pleased to hear that after sitting his examination three times he had at last managed to pass.

Adolf, those many years before, had tolerantly watched Konrad and his friend Niko earnestly and enthusiastically indulging their interest in geese and sticklebacks; nothing could then have been further from his thoughts than that the discoveries unfolding before his unseeing eyes would carry the two young men to far greater heights than he himself had achieved—even though his son had already firmly informed him of the importance of these studies for the eventual understanding of man himself. Adolf the raconteur might well have drawn a lesson from one of his own stories. Joining the Captain's table on an Atlantic cruise, he had seen the place-card for a Dr Schaudinn. In came the young, blond Estonian giant, and Adolf asked, 'Are you just an ordinary Schaudinn or *the* Schaudinn?' The stranger conceded he was indeed the Schaudinn who had at last discovered the spirochaete of syphilis, a microbe that had eluded thousands of trained eyes peering for decades into powerful microscopes; it was he who had spotted the brilliant and lively cork-screw creature the first time infected material was placed before him. How then, he was asked, had all these others failed to see it? 'That', replied Schaudinn, 'was because they were all physicians while I am a zoologist'. As a zoologist he had studied live specimens from the same family of microbes, but physicians had always stained their specimens, and the dye itself killed and destroyed the evidence they sought. Advances in one scientific discipline have often come by the extension of another, as Adolf knew.

A time when you specially miss a dead parent is when you have something you badly want to say to him, and in his imagination, Konrad put the news of his success to the memory of his father. Back came the vigorous, emphatic reply, '*Es ist unglaublich*—it is unbeliev-able—that rascal getting the Nobel Prize, and, of all things, for jack-daws!' Konrad, telling me this story, roared with delighted laughter, both with and on behalf of his father and Gretl joined in too. Adolf Lorenz would have glowed with pride for his son.

Besides the prestige that has increased year by year with the reflected glory of the growing band of Nobel Laureates, the actual cash value of the prize in 1973 was 510,000 Swedish crowns (some £50,000). With the eyes of his critics upon him, Lorenz at last spoke publicly to disown the pro-Nazi taint that had soured the end of his early prolific and produc-tive period. This was necessary, for the storm of criticism now reached its height. Shortly before Lorenz's arrival in Stockholm, Simon Wiesenthal, the veteran Jewish Nazi-hunter, had protested against the award of a Nobel Prize to the author of the offending 1940 paper, saying that Lorenz could not have failed to see what what was going on about him at that time, and the Swedish newspapers joined vigorously in the debate. In Stockholm, demonstrations were half expected, and a security man was set to guard Lorenz. An influential Swedish journalist who went to interview him expected to find a neo-Nazi, but was converted by Lorenz's charm and frankness, and finally wrote a favourable story. In the event, there were no 'incidents' to mar the visit.

Lorenz accepted the honour that was no less than his due, and firmly put his third share of the money into the drawer marked 'for scientific work': it would go towards the new aquarium at Altenberg. Gretl, the money manager, commented that she had taken not so much as a new dress from the award.

Lorenz planned a Nobel lecture based on the third chapter of the book he was writing: he would speak on the philosophical theory of values. But Professor Bengt Gustavsen, one of the judges, steered him away from this, suggesting that the earlier material on the value of analogy as a source of knowledge was more appropriate, and Gretl firmly agreed. Lorenz complied and a year later acknowledged to me that the advice had been right, for even in those twelve months his ideas on the philosophy of values had developed and matured; he would already have regretted a Nobel address so soon out of date.

Another smaller intended honour reopened the old Nazi sore. An old and conservative member of the Bavarian Academy of Science – Lorenz had known him slightly from the Academy sessions – approached to ask if he would accept an honour called 'The Schiller Prize'. Open and unsuspecting, Lorenz agreed and it was announced: he would receive the 'Schiller Prize of the German People' given for 'the German culture of the European spirit' – whatever that may be. To us it may sound innocuous enough, but 'Deutsches Kulturwerk Europäischen Geistes' speaks both sharp and sour to the German ear: rather late in the day Lorenz learned that the prize came from a neo-Nazi group. He had no reason to suspect it, he told me – after all, Schiller was no Nazi – but when at last he did look more closely he saw the spirit of Nazism still alive. He could have some sympathy with idealists who were taken in during the early days, but none at all with people who, knowing all the horrible, satanic things the Nazis did, could still, at this late hour, be neo-Nazis. He was horrified and disgusted.

It was convenient, perhaps, that he was able to claim the sanctuary of his sick-bed while his son Thomas and his friend Eibl-Eibesfeldt went to the ceremony to announce that the 10,000 deutschmarks (about £1,500) would be given to Amnesty International, a choice demonstrating Lorenz's belief in a broader political freedom that stands against the excesses of all regimes. This incident was also a last, late demonstration of his political naivety and need for protection against his own uncalculating responses. But just at that moment, following his greatest triumph, he had few of his old friends about him, and only just in time were the tables turned — although Amnesty did not in the event get the money, since those who had awarded the Schiller Prize failed to pay.

To his friends and those who see the great man in him (as do many younger scientists who may dispute his results or individual statements while finding inspiration in his leadership), Lorenz is in turn the subject of deep affection and wry exasperation. One pupil has said that no one really close to Lorenz can see him objectively. It can be difficult to follow him in changes that are less of opinion than the reflection of mood on opinion. He is well if selectively informed; he can seek out and listen to the views of those who have his respect and so adapts new thought into the body of his own ideas. But more often his conversation is transformed by sheer enthusiasm into a near-monologue that may be brilliant but disconcertingly difficult to respond to.

Seemingly arrogant and certainly assertive, he claims humility, and proclaims that a sense of humour is one of man's greatest assets since no one with a true sense of humour can be a megalomaniac, or can fail to be humble. If there is such a thing as a modern sense of humour, one that is more subtle than that of the past, this is a most promising element of man's cultural evolution. For this puzzling human capacity for humour evidently has an affinity with moral responsibility: it is a search for a sense of what is fitting, a mechanism to root out dishonesty.

As Lorenz approached seventy, I asked him how much work he saw still ahead of him. His eyes alight with the prospect of all he still wanted to do — among it the social and bonding studies with fish, geese and other creatures, the books to be written, including his second on the theory of knowledge and, at last, the popular book on geese — he replied: 'If as a scientist you are honest, your belief in the proportion of what you know is for ever changing for the worse . . . and I certainly have more work cut out for me than the few years given.' He paused and added, '. . . although I firmly intend to live as long as my father, who lived to ninety-two'.

Bibliography

Popular Books by Konrad Lorenz

King Solomon's Ring. London: Methuen, 1952; New York: Crowell, 1952. Translated from the German *Er redete mit dem Vieh, den Vögeln und den Fischen*. Vienna: Verlag Dr G. Borotha-Schoeler, 1949. The book that has introduced a generation of ethologists to their subject. The 1964 English edition has an introduction by W. H. Thorpe.

Man Meets Dog. London: Methuen, 1954; Boston: Houghton Mifflin, 1955. Translated from the German *So kam der Mensch auf den Hund*. Vienna: Verlag Dr G. Borotha-Schoeler, 1950. A common sense guide for dog (and cat) lovers. The theory of the evolution of domestic dogs that is advanced by Lorenz is disputed by other experts.

On Aggression. London: Methuen, 1966; New York: Harcourt, Brace & World, 1966. Translated from the German *Das sogenannte Böse: Zur Naturgeschichte der Agression*. Vienna: Verlag Dr G. Borotha-Schoeler, 1963. Detailed animal studies support Lorenz's thesis on the roles of aggressivity in animals. In the final, more controversial chapters, man is distinguished from other animals as the only species that preys on its own kind (which has since been disputed).

Civilised Man's Eight Deadly Sins. London: Methuen, 1974; New York: Harcourt Brace Jovanovich, 1974. A concise — perhaps overcompressed — view of the dangers to man that arise from his own nature and actions when natural selection is replaced by unnatural. Covers some territory familiar to environmentalists but from a peculiarly Lorenzian angle.

A book on Lorenz's work with geese is planned.

Scientific Works in English by Konrad Lorenz

Studies in Animal and Human Behaviour, Vol. I. London: Methuen, 1970; Cambridge (Mass.): Harvard University Press, 1970.
Contains his most important pre-war observational and theoretical papers.

1931 Contributions to the study of the ethology of social Corvidae
1932 A consideration of methods of identification of species-specific instinctive behaviour patterns in birds
1935 Companions as factors in the bird's environment
1937 The establishment of the instinct concept

Bibliography

1938   Taxis and instinctive behaviour pattern in egg-rolling by the Greylag goose

1942   Inductive and teleological psychology

Studies in Animal and Human Behaviour, Vol. II. London: Methuen, 1971; Cambridge (Mass.): Harvard University Press, 1971.
The post-war papers have a stronger philosophical, human and social content.

1941   Comparative studies of the motor patterns of Anatinae

1950   Part and parcel in animal and human societies

1954   Psychology and phylogeny

1958   Methods of approach to the problems of behaviour

1959   Gestalt perception as a source of scientific knowledge

1963   Do animals undergo subjective experience?

1963   A scientist's credo

Both volumes have an introduction in which Lorenz indicates how his ideas developed from one paper to the next. He also provides his own detailed critical annotations on these papers, and bibliographies.

Evolution and Modification of Behavior. Chicago: University of Chicago Press, 1965; London: Methuen, 1966.
Lorenz's own restatement of his position in relation to other schools of behavioural study, including 'the English-speaking ethologists'.

Behind the Mirror. London: Methuen; New York: Harcourt Brace Jovanovich, in press. On the biology of the theory of knowledge, with origins in his Königsberg-Kantian 'period and his writings as a prisoner of war. Much of it is well argued, whether or not the reader follows Lorenz all the way. (A companion volume to deal with the biological meaning of 'good' and 'evil' is planned.)

Konrad Lorenz: the Man and His Ideas (Compiled by Richard I. Evans). New York and London: Harcourt Brace Jovanovich, 1975. This book contains four papers by Lorenz:

1941   Kant's Doctrine of the A Priori in the Light of Contemporary Biology

1966   Evolution of Ritualisation in the Biological and Cultural Spheres

1970   The Enmity Between Generations and its probable Ethological Causes

1973   The Fashionable Fallacy of Dispensing with Description

(There is also a list of Lorenz's major works, quoting original rather than readily accessible sources; a conversation between Lorenz and Evans—not taken as source material for the present book; a critical essay by Donald Campbell and a reply by Lorenz.)

226

Some Other Scientific Papers and Lectures by Konrad Lorenz

These are in English except where stated. There are many more papers in German only.

1927 'Beobachtungen an Dohlen' (Observations on jackdaws). *Journal für Ornithologie*, 75, 511–19. In German only. From the diaries 'stolen' by his future wife and sent to Oskar Heinroth.

1934 'A contribution to the comparative sociology of colony-nesting birds'. *Proceedings of the VIIIth International Ornithological Congress*, London, 207–18.

1938 Über Ausfallserscheinungen im Instinktverhalten von Haustieren und ihre sozialpsychologische Bedeutung'.
(On the loss of instinctive behaviour of domestic animals and its importance for social psychology.) In *Kongress der Deutschen Gesellschaft für Psychologie in Bayreuth*. Leipzig: Johann Ambrosius Barth, 5,139–47. An early paper on this subject. Despite the time and place of its delivery the racial overtones are not strong – certainly when compared with the 1940 paper.

1939 'The comparative study of behaviour'. In Konrad Lorenz and Paul Leyhausen (Eds), *Motivation of Human and Animal Behaviour*. New York and London: Van Nostrand Reinhold, 1973. With Lorenz's paper are ten by Leyhausen in an orthodox Lorenzian style.

1940 'Durch Domestikation verursachte Störungen arteigenen Verhaltens' (Disorders caused by the domestication of species-specific behaviour). *Zeitschrift für angewandte Psychologie und Charakterkunde*, 59, 2–81. In German only. Nazi terminology is used in some passages, and Lorenz has been properly criticised for mixing this with science. But his theme, the danger of human genetic degeneration as a possible effect of self-domestication, is legitimate (if disputed by many geneticists), and Lorenz has frequently returned to it in more acceptable contexts.

1943 'Die angeborenen Formen möglicher Erfahrung' (The innate forms of possible experience). *Zeitschrift für Tierpsychologie*, 5, 235–409. In German only. Some parts were criticised by J. B. S. Haldane for their Nazi feeling; but the remainder is highly regarded by pupils (W. Schleidt, e.g.) as the best developed of Lorenz's papers dealing with 'releasers' and 'innate releasing mechanisms'.

1950 'The comparative method in studying innate behaviour patterns.' *Symposia of the Society for Experimental Biology*, 4, Animal Behaviour, Cambridge, 221–68.

1957 'The role of aggression in group formation.' *Transactions of the 4th Conference on Group Processes*. New York: Josiah Macy Jr Foundation.

Q

Bibliography

1962 'The function of colour in Coral Reef fishes.' *Proceedings of the Royal Institution of Great Britain*, 39, 282–6.

1969 'Innate bases of learning'. In Karl H. Pribram (Ed.), *On the Biology of Learning*. New York: Harcourt, Brace & World.

1973 'Analogy as a source of knowledge'. *Science*, 185, 229–34. This Nobel address and a scientific autobiography (in English) are included in Nobel Foundation, *Les Prix Nobel en 1973*. Amsterdam and New York: Elsevier, 1975.

Some Historical Precursors to Lorenz

Darwin, Charles. *The Expression of the Emotions in Man and Animals*. London: John Murray, 1872. The first major work on the evolution of behaviour.

Heinroth, Oskar. 'Beiträge zur Biologie, namentlich Ethologie und Psychologie der Anatiden'. *Verhandlungen des V internationalen Ornithologenkongresses*, Berlin, 1911, 589–702.

Huxley, Julian. 'The courtship-habits of the great crested grebe *(Podiceps cristatus)*'. *Proceedings of the Zoological Society of London*, 1914, 35, 491–562.

Whitman, Charles Otis. 'Animal Behaviour'. 16th lecture from *Biological Lectures of the Marine Biological Laboratories*. Woods Hole (Mass.), 1898.

Some Other Relevant Books and Papers

Ardrey, Robert. *The Territorial Imperative*. London: Collins, 1966; New York: Atheneum, 1966. A popular account of man as an aggressor. Appeared at about the same time as the English version of *On Aggression;* some reviews considered both together. Goes far beyond Lorenz: 'Handle carefully. Read with critical scepticism', wrote Geoffrey Gorer.

Bölsche, Wilhelm. *Die Schöpfungstage*. Dresden: Reissner, 1906. Lornez's own childhood introduction to Darwinian evolution.

Breland, Keller and Marian. *Animal Behavior*. New York: Macmillan, 1966. Intended as an introductory primer by a pupil of B. F. Skinner who found that innate behaviour could not, after all, be discounted.

Cloud, Wallace. 'Winners and sinners.' *The Sciences*, New York Academy of Science, December 1973, 13, 10, 16–21. A poorly researched article that has been quoted by opponents of Lorenz. (The journal now regrets an error in translation and those aspects of the article that amounted to no more than character assassination.)

Eibl-Eibesfeldt, Irenäus. *Ethology: The Biology of Behavior*. New York: Holt, Rinehart and Winston, 1970. The book that Lorenz had always intended to write.

Love and Hate. London: Methuen, 1972; New York: Holt, Rinehart and Winston, 1972. A popular account of human ethology.

Eisenberg, Leon. 'The *human* nature of human nature'. *Science,* 1972, 176, 123–8. Strongly attacks Lorenz's 1940 paper (which it mistranslates at one point) and, more generally, his extrapolation from animal to man.

Frisch, Karl von. *The Dancing Bees: An Account of the Life and Senses of the Honey Bee.* London: Methuen, 1954 (2nd edn 1966). Revised edition, New York: Harcourt Brace Jovanovich, 1966. An easy-to-read account by Lorenz's fellow winner of the Nobel Prize of the creature at the centre of his life's work.

Haldane, J. B. S. 'The argument from animals to men'. *Journal of the Royal Anthropological Institute,* 1956, 86, I, 1–14. Criticises Lorenz's 1943 paper.

Hinde, Robert A. 'Dichotomies in the study of development'. In J. M. Thoday and A. S. Parkes (Eds), *Genetic and Environmental Influences on Behaviour.* London: Oliver & Boyd, 1968. Describes how Lorenz's way of looking at animals differs from his own and others'.
Animal Behaviour. London and New York: McGraw-Hill, 1966 (2nd edn 1970). A good textbook that offers a synthesis of ethology and comparative psychology. This coming together of ethologists with such psychologists as the American Frank Beach and the Canadian Donald Hebb did much to define a new course for human behavioural studies.

Koenig, Otto. *Tales from the Vienna Woods.* London: Methuen, 1958. A popular view of animal life observed at the Wilhelminenberg Institute; with an introduction by Lorenz.

Lehrman, D. S. 'A critique of Konrad Lorenz's theory of instinctive behavior'. *Quarterly Review of Biology,* 1953, 28, 337–63. The classic American attack on Lorenz. Intelligent but overemphatic. Later Lehrman would have taken a more moderate stand.

Lorenz, Adolf. *My Life and Work: The Search for a Lost Glove.* New York and London: Charles Scribner's Sons, 1936. The vigorous, anecdotal account of his life by Konrad's father. Long out of print but delightfully easy to read. Written in English first, and later translated into his native German by the author.

Montagu, M. F. Ashley (Ed.). *Man and Aggression.* New York: Oxford University Press, 1968 (2nd edn 1973). Collected criticisms of Lorenz and Robert Ardrey (*The Territorial Imperative,* etc.).

Stade, George. 'Lorenz, and the dog beneath the skin'. *Hudson Review,* 1973, 26, 1, 60–86. Noteworthy for the fact that, entirely from Lorenz's writings in English, Stade draws an accurate picture of Lorenz and his aims. The exact opposite of the Cloud article (above).

Bibliography

Tiger, Lionel, and Fox, Robin. *The Imperial Animal*. New York: Holt, Rinehart and Winston, 1971; London: Secker and Warburg, 1972. Another animal-based view of man, with an introduction by Lorenz.

Tinbergen, Nikolaas. *The Study of Instinct*. Oxford: Clarendon Press of Oxford University Press, 1951. The classical text: a milestone in the development of ethology.

The Herring Gull's World. London: Collins, 1953. New York: Basic Books, 1961. A popular study of behaviour in a single, closely observed species.

Curious Naturalists. New York: Basic Books, 1958; London: Penguin, 1973. A popular account of Tinbergen 'the hunter' at work.

'On aims and methods of ethology'. *Zeitschrift für Tierpsychologie*, 1963, 20, 410–33. Attempts to bridge the gap between Lorenz's and other schools of ethology (that of Hinde, e.g.).

'On war and peace in animals and man'. *Science*, 1968, 160, 1411–18. Tinbergen on aggression.

'Ethology', a chapter in *Scientific Thought, 1900–1960* (Rom Harré, Ed.). Oxford: Clarendon Press of Oxford University Press, 1969, 238–68. A good historical exposition.

'Functional ethology and the human sciences'. *Proceedings of the Royal Society of London*, 1972, B. 182, 385–410. A Tinbergen's-eye view of some of civilised man's deadly sins, and how ethology contributes to their study.

'Ethology and stress diseases'. *Science*, 1974, 185, 20–7. This Nobel address also appeared in English in Nobel Foundation, *Les Prix Nobel en 1973*. Amsterdam and New York: Elsevier, 1975.

Wilson, E. O. *Sociobiology: The New Synthesis*. Cambridge (Mass.): The Belknap Press of Harvard University Press, 1975. Illustrates the post-Darwinian idea of 'inclusive fitness', which proposes that kinship plays a role in natural selection, since an individual's genes may be propagated not only by care of his own offspring, but also, to a lesser degree, by the care of his other family relations. Sets the scene for powerful biological explanations of aggression, altruism, etc. Like *On Aggression*, the end of the book extrapolates to man, and is controversial for the same reasons.

Index

Index